Towards Digital Enlightenment

Dirk Helbing

Editor

Towards Digital Enlightenment

Essays on the Dark and Light Sides
of the Digital Revolution

 Springer

Editor
Dirk Helbing
CLU E 1
ETH Zürich
Zürich, Switzerland

ISBN 978-3-319-90868-7   ISBN 978-3-319-90869-4   (eBook)
https://doi.org/10.1007/978-3-319-90869-4

Library of Congress Control Number: 2018951408

This Springer imprint is published by the registered company Springer Nature Switzerland AG
The registered company address is: Gewerbestrasse 11, 6330 Cham, Switzerland

Preface

When my book *Thinking Ahead* was published, our societies were on a path toward data dictatorship[1] or, as some people call it, technological totalitarianism.[2] Mass surveillance, as reported by Edward Snowden and others, was just the beginning. With the Digital Manifesto published in Spektrum der Wissenschaft,[3] it became increasingly clear that more and more areas of our lives were increasingly influenced by algorithms and often in such subtle ways that we had not even noticed this. Secretly, cookies used by our Internet browsers as well as our smartphones had delivered data to little known companies such as Axciom[4] or Axon Global,[5] which create detailed profiles about everyone living in the modern world—profiles that can predict our behavior better than our friends, family members, or even partners.[6] These profiles reveal more or less every relevant detail of our lives, including those that we consider highly private (such as the friends we have, our religious beliefs, sexual preferences, political inclination, and voting behavior). The company Crystal Knows[7] even runs a platform, which advertised for some time with the slogan "See anyone's

---

[1]Das DigitalManifest: Digitale Demokratie statt Datendiktatur, http://www.spektrum.de/pdf/digital-manifest/1376682; English translation: Will Democracy Survive Big Data and Artificial Intelligence? https://www.scientificamerican.com/article/will-democracy-survive-big-data-and-artificial-intelligence/

[2]F. Schirrmacher (ed.) Technologischer Totalitarismus: Eine Debatte (Suhrkamp, 2015) https://www.amazon.de/Technologischer-Totalitarismus-Eine-Debatte-suhrkamp/dp/3518074342; M. Schulz, Technologischer Totalitarismus: Warum wir jetzt kämpfen müssen (FAZ, 6.2.2014) http://www.faz.net/aktuell/feuilleton/debatten/die-digital-debatte/politik-in-der-digitalen-welt/technologischer-totalitarismus-warum-wir-jetzt-kaempfen-muessen-12786805.html

[3]See footnote 1.

[4]https://www.acxiom.com

[5]https://axoncyber.com

[6]Diese Firma weiss, was Sie denken (Tagesanzeiger, 3.12.2016) https://www.tagesanzeiger.ch/ausland/amerika/diese-firma-weiss-was-sie-denken/story/25805157

[7]https://www.crystalknows.com

personality." In fact, it allowed everybody to study the personality traits of other people—neighbors, friends, colleagues, and even strangers, competitors, and enemies.

Google and Facebook are the best-known companies that realized early on how personalized information can be used to influence our attention and, with this, our opinions, emotions, decision-making, and behavior.[8] One might call this mass mind manipulation, and secret services can do this, too (see the JTRIG program).

But that's not all. Apple started to collect our activity data and set up a partnership with IBM to let their Cognitive Computer "Watson" determine our health status and diseases.[9] Microsoft's Windows 10 was spying on users.[10] Amazon's Echo is listening our conversations.[11] Game stations are spying on us.[12] And Cambridge Analytica produces psychological profiles of hundreds of millions of citizens to manipulate, for example, people's voting behaviors. The company claims that the Brexit and the outcome of the US election were largely their success.[13] "Has the "atomic bomb" of the digital age exploded?", people started to ask. Or was it rather a "digital Fukushima"?[14]

Was Elon Musk right that artificial intelligence (AI) was potentially more dangerous than nuclear bombs—considering the fact that it was used for cyberattacks on critical infrastructures such as hospitals, nuclear power stations, the electricity grid, and the Internet? Are we "summoning the demon"?[15]

At least, it eventually becomes clear that we were heading toward a cybernetic society, in which algorithms increasingly control society and individual behaviors. By revealing our personality traits, we have become vulnerable to manipulation. This has been used not only by the marketing industry, but also by politics, where one speaks of "social engineering" and "nudging" (and when combined with Big Data, of "big nudging").[16] Probably since the Arab Spring, or even before, social bots were used to influence social media and, by this, people's opinions and

---

[8]R. Epstein, R.E. Robertson: The search engine manipulation effect (SEME) and its possible impact on the outcome of elections, PNAS 112, E4512-E4521 (2015); R.M. Bond et al. A 61-million-person experiment in social influence and political mobilization, Nature 489, 295-298 (2012); A.D. I.Kramer, J.E.Guillori, J.T.Hancock, Experimental evidence of massive-scale emotional contagion through social networks, PNAS 111, 8788-8790 (2014).

[9]IBM forms new health data analytics unit, extends Apple partnership, http://www.zdnet.com/article/ibm-forms-new-health-data-analytics-unit-extends-apple-partnership/

[10]Microsoft was just ordered to stop Windows 10 from spying on users, http://bgr.com/2016/07/22/microsoft-windows-10-data-collection-france/

[11]Alexa and Google Home record what you say. But what happens to that data? https://www.wired.com/2016/12/alexa-and-google-record-your-voice/

[12]Our game consoles are likely spying on us, and this is business as usual, https://www.polygon.com/2014/2/28/5456940/our-game-consoles-are-likely-spying-on-us-and-this-is-business-as

[13]Ich habe nur gezeigt, dass es die Bombe gibt, https://www.dasmagazin.ch/2016/12/03/ich-habe-nur-gezeigt-dass-es-die-bombe-gibt/

[14]Carsten Könneker, Fukushima der Künstlichen Intelligenz, http://www.spektrum.de/news/interview-die-unterschaetzten-risiken-der-kuenstlichen-intelligenz/1377620

[15]https://www.youtube.com/watch?v=_rfHNvHu8OE

[16]See footnote 1.

decisions. As algorithms determine news feeds, i.e., the number and kinds of people who receive particular messages, social media have become a powerful tool for propaganda and censorship, instruments of disinformation, and hybrid warfare.[17]

In the end, president Obama himself warned the public in his last Correspondents' dinner speech:[18] "... this is also a time around the world when some of the fundamental ideals of liberal democracies are under attack and when notions of objectively and of a free press and of facts and of evidence are trying to be undermined or in some cases ignored entirely. And in such a climate it's not enough just to give people a megaphone. And that's why your power and your responsibility to dig and to question and to counter distortions and untruths is more important than ... ever."

With my books *Thinking Ahead*[19] and *The Automation of Society Is Next*[20] as well as the various FuturICT Blogs and the science fiction *iGod*[21] (jointly written with Willemijn Dicke), I have tried to follow this request to dig deeper. I believe that this—together with the many dozens of talks I have given on digital issues to more than 10,000 people and the weekly news articles in various countries—has eventually had some impact on the international debate about where we should be heading in the digital age ahead of us. I have fundamentally challenged the idea of a data-driven "benevolent dictatorship" with the *Nature* article "Build Digital Democracy,"[22] the Digital Manifesto,[23] and my article "Why we need democracy 2.0 and capitalism 2.0 to survive."[24] Now, a new game is about to begin.[25] To get there, it was important to not only criticize mass surveillance and manipulation, but to come up with an alternative, better model for our future—a world in which capitalism and democracy would not fight with each other, but where they would be married together to unleash the benefits of both. Such a new socioeconomic framework, which could dramatically improve our future prospects in a world of limited material resources, can now be built by combining Internet of Things with Blockchain Technology and Complexity Science.[26]

---

[17]Notwehr against the Machine, http://www.zeit.de/digital/internet/2017-12/34c3-chaos-computer-club-kuenstliche-intelligenz

[18]The complete transcript of President Obama's 2016 White House correspondents' dinner speech, https://www.washingtonpost.com/news/reliable-source/wp/2016/05/01/the-complete-transcript-of-president-obamas-2016-white-house-correspondents-dinner-speech/

[19]https://www.amazon.com/Thinking-Ahead-Digital-Revolution-Participatory/dp/3319150774

[20]https://www.amazon.com/Automation-Society-Next-Survive-Revolution/dp/1518835414/

[21]https://www.amazon.com/iGod-Willemijn-Dicke/dp/1544271573/

[22]D. Helbing and E. Pournaras, Build Digital Democracy, Nature 527, 33-34 (2015) https://www.nature.com/news/society-build-digital-democracy-1.18690

[23]See footnote 1.

[24]D. Helbing, Why we need democracy 2.0 and capitalism 2.0 to survive, Jusletter IT (25 May 2016) https://www.researchgate.net/publication/301620104_Why_we_need_democracy_20_and_capitalism_20_to_survive

[25]D. Helbing, Digitization 2.0: A New Game Begins, https://www.researchgate.net/publication/317279118_Digitization_20_A_New_Game_Begins

[26]D. Helbing, An Urgent Appeal to Save the Planet, 7-part series in The Globalist, https://www.theglobalist.com/population-environment-technology-society-climate-change-disaster/

Particularly in its second part, this book offers an encouraging vision, how a better future could look like. In the past years, to those who were studying statistics and forecasts about our world, it appeared that we would eventually run into a situation of serious scarcity as well as economic and population collapse, as described by the Limits to Growth[27] and Global 2000[28] studies. Moreover, it seemed that we would lose our privacy, democracy, and human rights in an increasingly top-down controlled world trying to respond to these crises and disasters. But now there is a chance that we will take a different path into an alternative digital future characterized by peace and prosperity, which would be based on a digitally upgraded democracy supporting collective intelligence, jointly with a socio-ecological finance system ("finance 4.0") boosting the evolution of a circular and sharing economy or, in other words, a participatory information and innovation ecosystem.

Why am I getting more optimistic? Because an open letter has been calling for another kind of digital economy, recently.[29] Europe has come up with a Data Protection Directive. The highest European Court has forbidden today's kind of mass surveillance. Governments have started to worry about the dual use of artificially intelligent systems such as social bots. The White House has pushed for bottom-up approaches such as "Citizen Science" and "A Nation of Makers."[30] The IEEE has worked on guidelines for "Ethically Aligned Design,"[31] and leading IT companies have decided to collaborate in order to break the filter bubble[32] and develop account-able and moral AI systems, which "should be an extension of individual human wills and, in the spirit of liberty, as broadly and evenly distributed as possible."[33] An age of digital enlightenment, as it was called for by the Digital Manifesto,[34] seems to be on its way. It's now your turn to contribute this better future, which is just around the corner!

Zürich, Switzerland   Dirk Helbing
February 14, 2018

---

[27]D.H. Meadows, Limits to Growth (Signet, 1972).

[28]The Global 2000 Report to the President, http://www.geraldbarney.com/G2000Page.html

[29]Open Letter on the Digital Economy, https://www.technologyreview.com/s/538091/open-letter-on-the-digital-economy/; http://openletteronthedigitaleconomy.org

[30]https://obamawhitehouse.archives.gov/blog/2016/04/14/collaboration-gives-federal-govern ment-citizen-science-and-crowdsourcing-new-home; https://obamawhitehouse.archives.gov/node/316486

[31]Ethically Aligned Design, http://standards.ieee.org/develop/indconn/ec/ead_v1.pdf

[32]http://www.independent.co.uk/life-style/gadgets-and-tech/news/computer-simulation-world-matrix-scientists-elon-musk-artificial-intelligence-ai-a7347526.html; http://www.businessinsider.de/tech-billionaires-want-to-break-humans-out-of-a-computer-simulation-2016-10

[33]https://blog.openai.com/introducing-openai/

[34]See footnote 1.

Acknowledgments

The work presented in this book has benefited largely from the support by my teams and colleagues at ETH Zurich and TU Delft, from many discussions with people around the world, including the Complexity Science Hub Vienna, and from the following projects: ERC MOMENTUM (Advanced Investigator Grant No. 324247), CIMPLEX (EC Grant. No. 641191), SoBigData (EC grant no. 654024), ASSET (EC Grant No. 688364), and FuturICT 2.0 (ERANET/SNF grant no. 20FE-1_170226). Thank you all!

Contents

Chapter 1
The World Today: A Net Assessment

Dirk Helbing

Is it mere alarmism to talk about global emergency? Or do we really have reasons to be alarmed?

"Time to evacuation" is the time between the ringing of an alarm, say, in a cinema or theater, and the time when the first people begin to evacuate. This time span is decisive for the number of people who will survive a building fire.

When the alarm rings, many people will not flee right away. After all, it could be a false alarm. Is there any sign the situation is serious? Is there any smoke? Does it smell strange?

If not, what are the other people doing? Are they evacuating themselves? Then it might be better to join the flow of people trying to get out. And yet, often nobody wants to be first. It might be a false alarm, after all.

Also, what if it isn't any safer outside? In such a case, shouldn't we ignore the alarm altogether, for lack of reasonable alternatives?

This article by Dirk Helbing was first published in the Globalist on September 12, 2017, under the URL https://www.theglobalist.com/technology-society-sustainability-future-humanity/ (reprinted with permission).

D. Helbing (✉)
ETH Zurich, Zürich, Switzerland

TU Delft, Delft, Netherlands

Complexity Science Hub, Vienna, Austria
e-mail: dhelbing@ethz.ch

1.1 Planet Earth Is in Trouble

This is the kind of situation humanity currently finds itself in. Planet Earth is in trouble. I admit, we have heard that many times—and it often seemed to be a false alarm. So, we have to get a clearer sense of how serious the situation really is.

Today's economy is not sustainable, we hear. The world is overconsuming the Earth's renewable resources by 50%, Europe by 250% and the United States by 400%, we read.

In less abstract terms, that means nothing else than that "industrialized countries are living at the cost of developing countries" and that "today's generation is living at the cost of future generations."

It's not fair, but it's the way things have always been, one might say. The powerful exploit the weak—not nice, but a fact of life.

So far, we may have disliked such exploitation, but people in today's industrialized societies have implicitly consented to their political and economic system, which produces such drastic inequality in wealth and opportunities.

Most likely, their reason for acquiescence was that they did not have much to complain. Performance indicators have been growing for years: The world's gross domestic product (GDP) went up. The number of wars went down. Average standards of living increased. And so on.

1.2 Misleading Indicators

However, a closer look is revealing. The Millennium Goals were not met. Most of the progress that was made occurred mainly thanks to China's enormous development, while many world regions are still crumbling.

Moreover, indices measuring unemployment or GDP per capita have been conveniently—but usually silently—changed many times. What used to count as unemployed has, in part, been replaced by so-called one Euro jobs (or internships or other precarious kinds of employment).

Similarly, GDP per capita is manipulated by changing the composition of goods people are assumed to consume. For example, if a regular steak become too expensive for many to afford, it may be replaced by cheaper goods rather than counting as inflation. This artificially keeps inflation down and economic growth rates up.

In other words, "most jobs ever" and "lowest inflation ever" is often a result of manipulation—and that, in turn, is a sign of how serious the situation really is.

1.3   Fallout from the Financial Crisis

Of course, the fallout from the financial crisis still looms large—larger than most people realize. The relevant fact to keep in mind is that, so far, the upper classes of society did not suffer much from it. Central bankers came to their help.

However, public debt has increased dramatically, and this is basically everyone's debt. Thus, aside from the fact that we are already consuming more than we can presently pay, most people would not be able to ever pay for their share of the public debt, and those who could will make sure through the political process that they are protected.

As a result, there is no prospect that most governments will ever be able or willing to pay off their debts.

This debt is concentrated in industrialized countries, which reflects again that the industrialized nations' style of living is neither justified nor sustainable. It comes at the cost of our future, living on the backs of coming generations, and on the backs of other peoples.

Pay day is coming closer. Mass migration and terrorism are challenging the old powers. Forecasts by the Club of Rome and others imply that there are billions of people too much on this planet—commonly referred to as "over-population."

We cannot keep the problems caused by us at bay much longer—if we don't change the system.

1.4   Talk of World War III

Never before have people talked so much about a looming World War III. That is not really an option, however, as it would make the Northern hemisphere basically uninhabitable.

No wonder, though, that business for nuclear-proof bunkers for the elites and ranches in far-away places in the Southern hemisphere for the same elites are booming. They know things can't go on like this much longer.

One thing is clear: Continuing with today's economic system will surely end in disaster. In the past, we have perhaps not talked and read so much about this situation—better alternatives were lacking. The approach was to enjoy life as long as possible, if we could not do much about our existential threats.

1.5   Time to Start a New Chapter of Human History

Now, however, it's time to shout "the world is on fire." We can "evacuate ourselves" from a future that is doomed, because there are new concepts and technologies that allows us to build a different, better future.

Artificial Intelligence, the Internet of Things, and Blockchain Technology, 3D printers, and the sharing economy, for example, are potential game changers. They allow for a better organization of the world.

A new game begins. A better coordination of resources, empowered by digital technologies, will enable a better monetary, financial and economic system, and political systems can be upgraded as well, as Chap. 10 lays out.

In other words, if we act now, we aren't doomed, but major changes are needed. It's time to take our future in our hands and start a new chapter of human history, as it happened when the industrial revolution was on its way.

Metaphorically, we may imagine it like the transformation from a caterpillar to a butterfly. In other words, the outcome may be much nicer than the future we imagine today.

1.6 Takeaways

- When the alarm rings, many people will not flee right away. After all, it could be a false alarm.
- The world is overconsuming the Earth's renewable resources by 50%, Europe by 250% and the United States by 400%.
- The fallout from the financial crisis still looms large—larger than most people realize.
- We cannot keep the problems caused by us at bay much longer—if we don't change the system.
- One thing is clear: Continuing with today's economic system will surely end in disaster.

Chapter 2
Why Our Innovation System Is Failing and How to Change This

Dirk Helbing

Our innovation system has terribly failed. It is well designed to support gradual improvements of our knowledge and technologies. But it does not support disruptive innovations well, which would create new qualities and functionalities, or question the basis of our established knowledge and routines. Moreover, our knowledge does not keep up anymore with the pace at which our world changes, and solutions to new problems often come with serious delays. Therefore, we need to re-invent innovation. In particularly, we must learn to create systems embracing collective intelligence that surpasses the intelligence of even the brightest individual and of powerful supercomputing solutions. This cannot be based on top-down nor majority decisions. Diversity is absolutely crucial for collective intelligence to work...

2.1 The Innovation Crisis

In times of economic recessions and political crises, innovations and new ideas are bitterly needed. But great ideas are rare, and many ideas that appear to be new just reproduce or re-invent what somebody else has thought before. As I will show below, it is very difficult to have just one or two great ideas in a lifetime that will survive for more than 50 years, or even change the world. This is bad, given today's world is changing faster than ever due to climate change, environmental change,

This chapter by Dirk Helbing was first published as a FuturICT blog on August 7, 2016, under the URL http://futurict.blogspot.com.eg/2016/08/why-our-innovation-system-is-failing_87.html

D. Helbing (✉)
ETH Zurich, Zürich, Switzerland

TU Delft, Delft, Netherlands

Complexity Science Hub, Vienna, Austria
e-mail: dhelbing@ethz.ch

© Springer International Publishing AG, part of Springer Nature 2019
D. Helbing (ed.), *Towards Digital Enlightenment*,
https://doi.org/10.1007/978-3-319-90869-4_2

demographic change, conflict, war etc. Do we innovate quickly enough? Does our knowledge still keep up with this rapid pace of change? I don't think so.

I certainly don't deny that the digital revolution is brewing a perfect storm. Within just a few years, we have seen many new technologies such as social media, Big Data, cloud computing, Artificial Intelligence, cognitive computing, Internet of Things, Blockchain technology as well as virtual and augmented reality—and I love them. But which of these inventions will create anything that will remain for 5000 years, or just 50? "Creative destruction" is often cited as ideal, because a new and better world order would be born from the chaos created. However, if we look back at the events so far, chaos has mainly born more chaos, in a gigantic global cascading effect that poses existential threats. Think of the events after September 11, 2001, or the financial, economic and public spending crisis, which hit us pretty unprepared.

In 2013, The Economist started "The great innovation debate".[1] It raised the question, whether we will ever invent something as important as the toilet again. It is often said that Big Data is the "oil of the twenty first century", but people increasingly add that, apparently, we haven't invented the motor yet to use it. Or if we have invented it, we have failed to build it due to political or economic constraints. Let me stress that I love "moonshot projects", including many of Google's efforts in this direction. Some people believe that, thanks to superintelligent systems, we will have all problems of the world solved by 2036—apart from climate change.[2] However, this seriously underestimates the nature of complex systems such as our society and economy. For sure, we will be faced with new challenges such as cyber threats. And so far, none of our attempts have been able to restart the engine of the world economy, or create prosperity for all—a fact that has been criticized by an Open Letter on the Digital Economy, which obtained a broad support.[3] In the last decades, inequality has further increased. In times of stagnation, this means that the lower and middle class, in other words: the small and medium sized companies had to pay for this. The diverse "ecosystem of customers and firms" that makes up a thriving economy[4] has increasingly degraded. It trends towards the creation of monopolies, particularly in the IT industry. While these monopolies increasingly claim global leadership where our societies should head, I don't see that they would have the recipes to create global well-being.

According to mainstream economics, a lack of innovations should never occur. If a problem would just grow big enough, they argue, there would be an increasing willingness to pay a high price for a solution. Accordingly, scientists and engineers would be incentivized to work hard to find a solution. Therefore, any big enough problem would be fixed sooner or later. However, even though the fact of global

---

[1]http://www.economist.com/news/leaders/21569393-fears-innovation-slowing-are-exaggerated-governments-need-help-it-along-great

[2]and cybercrime, that's what I learned from Jim Spohrer of IBM.

[3]http://openletteronthedigitaleconomy.org/

[4]http://science.sciencemag.org/content/317/5837/482.short

warming due to our carbon-based economy has been known already in the 1960s,[5] 50 years later we are still lacking a solution to the problem. Climate change may erase as much as one sixth of all species, and it poses an existential threat to humanity. By signing the Paris Climate Agreements, it has finally been admitted that innovations did not manage to solve this problem, even though many billions have been invested.

I personally believe those innovations haven't gone deep enough. We need to think out of the box, but we have stayed within it. As Albert Einstein said, we can't solve problems with the same kind of thinking that created them. Let me illustrate this for the example of the future prospects of our world. Back in the 1970s, The Club of Rome published their Limits to Growth study. No matter how hard they tried, they could not find a sustainable development path for the Earth. We would, therefore, run into economic and population collapse. Even though the study was highly controversial, the Global 2000 report commissioned by president Bill Clinton came to very similar results. So, if we want to avoid economic and population collapse, we must change the system of equations underlying the simulation scenarios. This means nothing else than the need to change our socio-economic framework. That is exactly what "finance 4.0", a new socio-ecological financial system could do—by measuring, valuating, and trading externalities and creating a multi-dimensional incentive system that would boost a circular and sharing economy by unleashing powerful market forces.[6] It is time that responsible economic and political players take action on this (see Chap. 10 in this book).

And what about our health system? In fact, the pharmaceutical industry is in big trouble, too. Most of the relevant companies have a declining number of new drug registrations. Even though the large multi-national companies are buying lots of startups to stay on top of innovation, this hasn't changed their situation much. In the meantime, we are running out of antibiotics that are effective against multi-resistant strains of bacteria. This is a serious issue, indicating a failure of our current research and development approach.

Some people are even more pessimistic than that. They claim that there haven't been any great innovations since Darwin's theory of evolution and Einstein's theory of relativity. Unfortunately, this is probably true, and it has reasons. Science is increasingly run like a business, measured by performance indicators. But while we perform better and better according to these indicators and despite the highest number of publications and patents ever, many of the problems our society is facing haven't been fixed. We are far from having the answers that our quickly changing world demands from us. For example, many people think we are far behind the Millennium goals or the United Nation's 2050 development goals.

---

[5]https://insideclimatenews.org/news/13042016/climate-change-global-warming-oil-industry-radar-1960s-exxon-api-co2-fossil-fuels

[6]D. Helbing, A Digital World to Thrive in: http://bit.ly/20T9BpX; Why We Need Democracy 2.0 and Capitalism 2.0 to Survive: http://bit.ly/1O5axWZ; Society 4.0: Upgrading society, but how? https://www.researchgate.net/publication/304352735

Unfortunately, the need to accelerate innovation will increase even more. According to Moore's law, computational power is doubling every 18 months. In 10–20 years, supercomputers are expected to exceed the processing capacity of a human brain. We will have computer programs capable of teaching themselves, robots producing other robots, and they will quickly improve over time. Hence, computer algorithms and robots will take over many of today's jobs. Anything that follows certain procedures, probably around 50% of all jobs in the industrial and services sector, could be performed by them cheaper and better. How can we cope with this challenge of having to reinvent half of our economy in just two decades? And how can we adapt our social, economic and legal system over such a short time? We need an Innovation Accelerator. But how would it work?

2.2   An Outdated Innovation System

Let us first analyze how we innovate today. While I will focus on the academic system, I expect that similar problems occur in industrial innovation systems as well. Currently, innovation is mainly happening in a competitive way. Each scientist, each company competes against all the others. Such innovation is expensive, slow, costly, and duplicates many results. This has led to an increasing percentage of programmatic research, which works roughly as follows: First, a ministry or agency determines research needs, emerging trends and knowledge gaps. It probably takes a few years until these become obvious, and it also takes some time to mobilize the budget (see, for example, the European Union's famous 7 year plans). But once a problem has been identified and the budget set aside, a call for proposals is launched.

These calls are usually oversubscribed, because there are never enough resources for everyone. This problem is "solved" by so-called "scientific beauty contests." Basically, one puts as many obstacles in the way as needed to retain the number of proposals that can be funded, i.e. one makes proposal-writing a complicated and time-consuming task. Selecting proposals is an equally complicated task. It is usually based on peer review—a process that consumes an increasing fraction of time as well. In the meantime, scientists probably spend about 40% of their time on proposal writing, reporting and reviews. Obviously, this time and money is lost for research, but the reason given for the inefficient administration is "having to justify how tax payers' money is spent".

2.3   Forget About Determining the Best Innovations Beforehand!

Can we at least be sure that the best contributions are selected? This is hard to say, as most non-funded projects are never carried out. However, the same peer review process is applied by scientific journals to select manuscripts for publication. Since many manuscripts that are rejected in the originally chosen journal are eventually published in a lower-ranked journal, we know a bit more about the quality of rejected as compared to accepted manuscripts. Surprisingly it turns out that a significant percentage of rejected manuscripts performs better than the ones that were accepted in the journal of choice.[7] A further surprise is that the majority of accepted and published papers performs significantly below the average of the journal, because a few of its publications generate most of its impact. This basically shows that quality is very difficult to judge. Quality may become obvious only over a long time. In fact, the recommendations of referees are often extremely divergent, particularly for innovative contributions, for which established standards do not exist.

But let's assume the best proposals were selected after a typically half-year-long review process. Then, one must find suitable staff to work on the project. The working contracts will start about 6 months after the acceptance of the proposal. The project will typically take 3 or 4 years, basically until a PhD is obtained. Publication of the research results will require between 6 months and 3 years, depending on the research field and journal. In any case, it's safe to say that this is not a fast process, and that it may easily take 10 years between the emergence of a new problem and its solution. If such a solution finally enters the knowledge core of the field (i.e. when it enters educational books and programs, which is the exception rather than the rule), it will take another 10–30 years, until it becomes best practice in business and administration. In cases of commercial industrial bias, scientific progress is often delayed by at least another 20 years, as it happened in the tobacco and energy sectors.[8]

This situation is reinforced by making science increasingly dependent on industrial funding—typically in order to fix today's problems rather than thinking ahead to find new approaches for the future. The only exception are projects of strategic nature that are of national importance. Many of these are aimed at accumulating power. But power is not the solution to many of our problems in a highly networked world, where strong interference will have unexpected side effects, feedback effects and cascading effects and often destroy structures that are essential for our society and economy to function.

Therefore, my personal judgment is that, on the one hand, our research and development (R&D) system has become increasingly dysfunctional. On the other hand, I know that many of the most successful publications are produced under difficult circumstances—they result from spontaneous ideas that someone decides to follow in the spare time, even though there is basically no funding for this. I have often wondered, why this is the case. To get funding is not the biggest problem. The problem is that we are asked to plan innovations, while the best ideas just happen. Funding institutions love well-elaborated proposals, but once one can elaborate new ideas in detail, they are not anymore cutting edge.

---

[7]In fact, a number of Nobel-prize winning discoveries have been rejected by scientific journals, or published in low-level journals.
[8]https://www.smokeandfumes.org/documents/document16, https://insideclimatenews.org/news/13042016/climate-change-global-warming-oil-industry-radar-1960s-exxon-api-co2-fossil-fuels

Great ideas must be pursued immediately, without the delays imposed by conventional funding mechanisms. However, our current innovation system makes scientists spend their time on ideas they get money for, and usually there is no time left for others. This effectively undermines academic freedom, and in many cases, certain innovations will be delayed for years, or even buried forever. By the time money becomes available, there will be other exciting ideas.

Thus, the crucial question is how to produce results more or less in real-time? Given that, today, it takes about 30 years from an invention to the real-world application of ideas, and many good ideas will never make it—how can we shorten this process to 5 years, or even 5 months or 5 weeks? And how can we increase the success rate of inventions? If we understood this, we could produce a dramatic innovation boost. Given that we have already more than 20 million unemployed people in Europe alone, such an innovation boost would be bitterly needed.

2.4   Everyone Wants Innovations, but Opposes Them!

To improve the innovation mechanism, it is important to first understand the nature of innovation a bit better. It turns out that there are two different kinds of innovations: gradual and disruptive innovations. Gradual innovations can be measured according to established standards of a field. They may best be characterized as "improvements"—such as a motor that consumes less energy and produces less emissions. Here, it seems reasonable to expect consensus of the reviewers in a funding board.

Pioneering research, in contrast, produces disruptive innovations, exploring or creating entirely new quality dimensions. These are the "true" innovations, which decision-makers and business people are usually keen on. Thus, why don't they happen more often? Almost by definition, disruptive innovations can't be assessed with established standards. They transcend existing categories and require one to think "out of the box." Consequently, such innovations are often highly controversial, and majority decisions of the reviewers in a funding board will rarely support them.

History shows that basically every disruptive innovation has been opposed in the beginning. The following quotes are quite illustrative. Ten years after the first successful test of electric light bulbs on October 22, 1879, Thomas Edison said: "Fooling around with alternating current is just a waste of time. Nobody will use it, ever." But today, everyone is using this kind of electricity 24 hours, 7 days a week. Or take the US president Rutherford B. Hayes. After a demonstration of Alexander Bell's telephone in 1872, he concluded: "It's a great invention but who would want to use it anyway?" Later, the inventor Lee De Forest (1873–1961) stated: "While theoretically and technically television may be feasible, commercially and financially it is an impossibility." Similar opinions were voiced, when the radio, planes, drilling for oil, or nuclear energy production were proposed.

It's no wonder that Alexander von Humboldt (1769–1859), one of the great discoverers of the world (and inventor of our modern university system) came to conclude: First, people deny that the innovation is required. Then, people deny that the innovation is effective. Afterwards, people deny that the innovation is important, and it will justify the effort to adopt it. Finally, people accept and adopt the innovation, enjoy its benefits, attribute it to people other than the innovator, and deny the existence of the previous stages. In other words: most innovations won't make it.

The famous quantum physicist Max Planck (1858–1947) even claimed: "Science advances one funeral at a time." This is mainly a result of the "rich gets richer effect," as Robert Merton (1910–2003) called it: while new inventions are made all the time, highly referenced work tends to get an even increasing amount of attention. This creates a threshold effect: only ideas that manage to get above the attention threshold will have a chance to win through. Such ideas are called "revolutionary ideas" and cause sudden, fundamental changes of our understanding or even of our world, so-called paradigm shifts, as analyzed by science historian Thomas Kuhn (1922–1996).

Revolutionary breakthroughs trigger an avalanche of new ideas, change the perspective of our world, and have the potential to transform our reality. One spectacular example is the replacement of the human-centered ("geocentric") world-view assuming the Earth to be the center of the universe by our current view that our planets would circle around the sun ("heliocentric" view). This shift goes back to observations of Nicolaus Copernicus (1473–1543) and theoretical work of Galileo Galilei (1564–1642). Later, it allowed Isaac Newton (1642–1726) to come up with his equations for the dynamics of celestial bodies. Without these discoveries, we would not be able to send rockets to the moon or have satellites circle around the earth. What seems to be a natural point of view today questioned the Christian worldview so much that Galilei was sent to prison. Only 350 years later, the Catholic church apologized for this.

The discoveries of Charles Darwin (1809–1882) were not less shocking. His theory of evolution—implying that humans were descendants from apes—largely replaced the idea of divine creationism, and is still questioned by some people today. However, without this paradigm change, it would be hard to imagine genetic engineering today. Or think of the theory of relativity by Albert Einstein (1879–1955). Without it, we would not understand how to produce nuclear energy or how to operate the Global Positioning System (GPS) exactly. In 1931, a provocative book entitled "100 Authors Against Einstein" even tried to discredit his work. Nevertheless, after a couple of decades, Einstein's counterintuitive predictions were finally confirmed. Such revolutionary ideas are extreme events—one might even say: "black swans",[9] i.e. unexpected rare events that cause major shifts (often through a massive cascading effect). They occur only every fifty or hundred years or so.

---

[9]N.N. Taleb, The Black Swan (Random House, 2nd edition, 2010).

But if great ideas cannot be identified beforehand, why don't we engage in refunding excellent work that has already happened, rather than in funding people for impressive promises made in lengthy and complicated project proposals? In other words, why do we pay money for the best promises, and not the best results? Wouldn't it save billions of tax payers' money, if the relatively few brilliant minds that exist could concentrate on innovations rather than on proposal writing, evaluation, and reporting? The good point about the scarcity of ground-breaking ideas is that funding agencies would have more than enough money to (re-)fund them. The open problem though is how to identify them (which, as I said before, cannot be well done by consensus or majority decisions of funding boards, as long as the ideas are young and quality criteria and research communities are not yet established).

2.5 Detecting Game-Changing Ideas and the Innovators Behind Them

But there is, in fact, a way of detecting where new ideas are produced, and where they are consumed. I did such a study together with Amin Mazloumian, Katy Börner, and others.[10] Each scientific publication refers to others it has been inspired by. Therefore, one can identify the flow of ideas in the world, and what are the places that produce ideas that are over-proportionally successful (shown in green in Fig. 2.1).

It is even possible to reveal what are the main ideas discussed in these publications—by analyzing the spreading of "memes." Memes are single words or combinations of words that appear in texts such as scientific publications. In physics, "atom" or "quantum mechanics" or "high-temperature superconductivity" would be such examples. In fact, my postdoc Tobias Kuhn and I, together with Matjaz Perc, have scanned the abstracts of all publications of the American Physical Society for such memes. This allowed us to identify the first occurrences of any new word or combination of words (meme). Of course, scientists come up with new terms all the time, but only in a few cases is the usage frequency quickly growing significantly in time. If this is the case, a new trend is born. Moreover, there is another property that characterizes Earth-shaking ideas. Their memes are "inherited" through the citation graph, i.e. they spread through mentions in later publications of colleagues. This separates important scientific concepts from meaningless memes. In fact, the history of the most important fields in physics can be determined in a fully automated way[11] (see Fig. 2.2).

---

[10]A. Mazloumian, D. Helbing, S. Lozano, R. P. Light and K. Börner (2013) http://www.nature.com/srep/2013/130130/srep01167/full/srep01167.htmlGlobal multi-level analysis of the 'Scientific Food Web'. http://www.nature.com/srep/index.html Scientific Reports 3, 1167.

[11]T. Kuhn, M. Perc, and D. Helbing (2014) Inheritance patterns in citation networks reveal scientific memes. Physical Review X 4, 041036, https://journals.aps.org/prx/abstract/10.1103/PhysRevX.4.041036

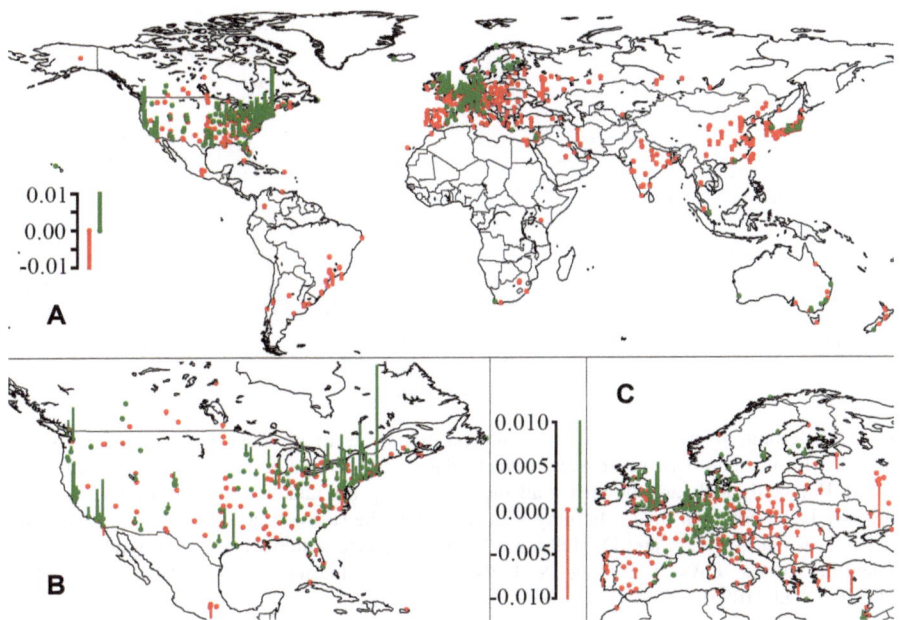

Fig. 2.1 Distribution of the birth of new, successful scientific ideas (**a**) in the world, (**b**) in the USA, (c) in Europe (From A. Mazloumian, D. Helbing, S. Lozano, R.P. Light and K. Börner, Global multi-level analysis of the 'Scientific Food Web'. *Scientific Reports* **3**: 1167 (2013), https://www.nature.com/articles/srep01167). Green bars indicate that the number of citations received is over-proportional, red that the number of citations received is lower than expected (according to a homogeneous distribution of citations over all cities that have published more than 500 papers)

Note, however, that data mining of publications and citations cannot only identify new trends, but also the key researchers in the field. Together with Amin Mazloumian, Santo Fortunato and others, I found that a milestone paper is one that doesn't only attract many citations, but also draws more attention to the entire body of work of a researcher. Consequently, many of his or her previously published papers will be cited more frequently. In other words, a milestone paper boosts other papers of the same author, which can easily be measured.[12] This fact allows one to identify rising stars at the firmament of science, which usually stay on top of their respective fields in terms of citations for their entire careers. In fact, all Nobel prize winners have such a milestone paper, which boosted the visibility of their whole body of work and, thereby, their entire career. But there are only a few scientists, who have two or more big boosts in their careers, i.e. two or more milestone papers (see Fig. 2.3). Hence, it is really affordable for a research institution to flood the few researchers, who succeeded in having a relevant boo effect, with money.

---

[12] A. Mazloumian, Y-H. Eom, D. Helbing, S. Lozano, and S. Fortunato (2011) http://www.plosone.org/article/info%3Adoi%2F10.1371%2Fjournal.pone.0018975 How citation boosts promote scientific paradigm shifts and Nobel Prizes. http://www.plosone.org/ PLoS ONE 6(5), e18975.

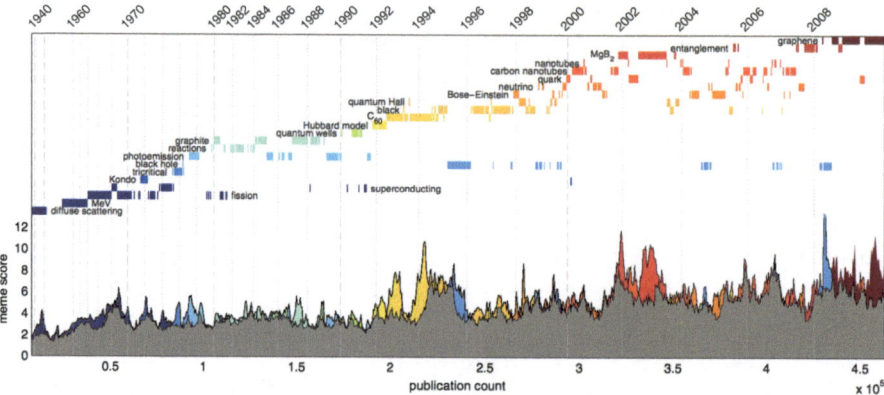

Fig. 2.2 Evolution of the most successful research fields in physics, based on meme scores obtained from the American Physical Society data set. The time axis is scaled by publication count. Bars and labels are shown for all memes that top the rankings for at least 10 out of the displayed 911 points in time. The gray area represents the second-ranked meme at a given time (from T. Kuhn, M. Perc, and D. Helbing (2014) Inheritance patterns in citation networks reveal scientific memes. Physical Review X 4, 041036, https://journals.aps.org/prx/abstract/10.1103/PhysRevX.4.041036)

2.6   Chance: The True Father of New Ideas?

However, the perhaps most surprising implication of the above discussion is that it is often impossible to determine in advance how successful an idea or someone will be. Scientists often have difficulties to judge themselves, which of their papers will be most successful. And this reveals another important feature of the nature of innovation. A successful innovation requires at least two things: First, an invention must be made, and second, it must spread. We often believe that great inventions will result, if well-skilled experts undertake a large enough and systematic effort. However, they rarely occur without the contribution of chance. You may call it "luck" or "serendipity"—naming it "creativity" certainly sounds more deserving. In fact, many great inventions are based on trial and error, and generating suitable settings, which allow randomness to contribute, is an art. For example, the famous inventor and electrical pioneer Thomas Alva Edison (1847–1931) noted: "I have not failed. I've just found 10,000 ways that won't work." Moreover, even people who made a great invention, often do not recognize its potential. For example, James Clerk Maxwell (1831–1879), who formulated the theory of electricity once said: "I do not know what electricity is good for, but I'm pretty sure that Her Majesty's government will soon tax it." How right he was! But practical applications spread much later. The same was true for the MASER (the predecessor of the LASER).

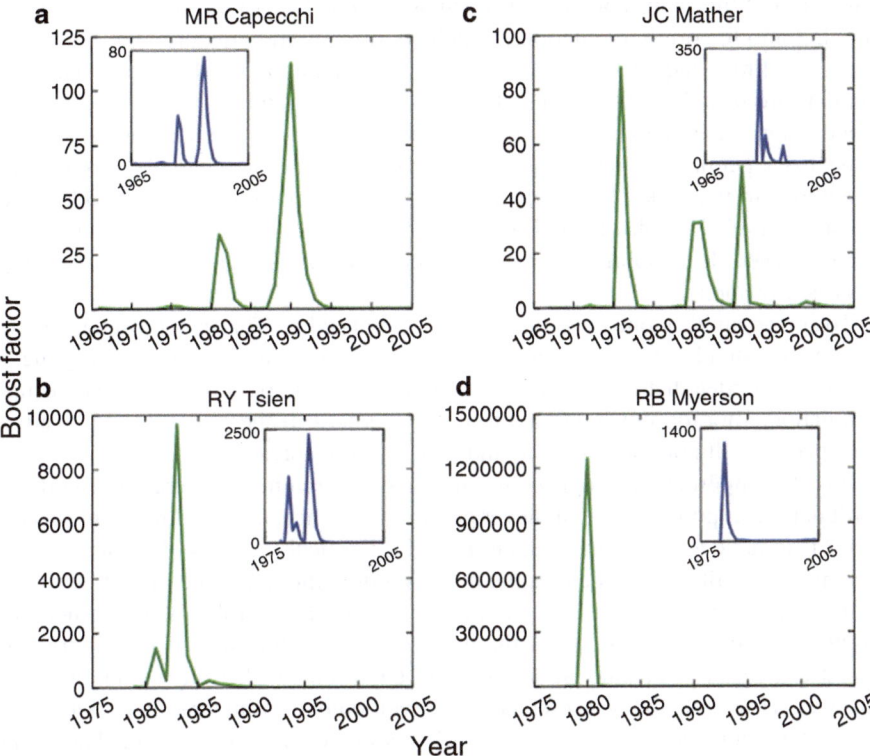

Fig. 2.3 Temporal evolution of boost factors for Nobel Laureates [here for (**a**) Mario R. Capecchi (Medicine, 2007), (**b**) John C. Mather (Physics, 2006), (**c**) Roger Y. Tsien (Chemistry, 2008) and (**d**) Roger B. Myerson (Economics, 2007)] (A. Mazloumian, Y.-H. Eom, D. Helbing, S. Lozano, and S. Fortunato, How citation boosts promote paradigm shifts and Nobel Prizes, PLoS ONE 6(5): e18975. https://doi.org/10.1371/journal.pone.0018975.). Sharp peaks indicate citation boosts in favor of older papers, triggered by the publication and recognition of a landmark paper. Insets: The peaks even persist (though somewhat smaller), if in the determination of the citation counts, the landmark paper is skipped

When the invention was made, it appeared pretty useless, but in the meantime, a huge market emerged around LASERs.

2.7   Social Factors Matter

The idea that a successful innovator is a lonely genius is often misleading. Successful innovations are also the result of social processes. The ability of innovations to spread is a crucial aspect. In fact, the "ignored genius" has become a well-known proverb. Sometimes, ideas are born ahead of time. But in order to spread, they must come at the right moment. The context must be fitting. It's helpful if many people have already been waiting for the idea to occur. So, what determines whether an idea comes just at the right moment? This actually depends a lot on the social network. Looking into the origin of citations, which are often used to quantify the success of ideas, my PhD student Christian Schulz made an important discovery: The first citations are usually self-citations, then citations by co-authors tend to follow, and later citations of co-authors of co-authors. Only less than 40% of citations on average come from people not related to the inventor directly or indirectly through the co-authorship network. Hence, one may interpret the co-authorship network as the "infrastructure" through which ideas spread. In other words, an isolated inventor will rarely succeed. It's important to have a social network through which the ideas can successfully spread.

To spread an idea, one must furthermore "get it right." This is not just a matter of formulating the idea correctly—it's also a matter of presentation, i.e. the storyline or narrative. An idea that fits current trends, but creates the right degree of intellectual stimulation—not too much and not too little—is best. If the idea is too innovative, most people will not understand it, and so it will not spread.

It is also important to recognize that an initial idea is almost never right. But if you don't dare to expose it, you will never get it right. Hence, another important success factor is to have a social environment that gives critical feedback. I often say that I owe my competitors and enemies the most. While critique is annoying, it helps one to frame an idea well, and it often reveals the unsolved problems that others are struggling with, i.e. the solutions the world is waiting for. Why is this important? Because it means that innovations need an environment that allows one to make mistakes. However, today's research and development settings rarely provide such an environment, and scientific consensus is favored over controversial debates. The Silicon Valley is rather the exception than the rule. It's known to allow people to experiment and make mistakes. Those who fail, are not singled out. They will find venture capital and another job again, because they are considered to be people, who have (probably) learned a lesson. Some people think, this particular culture has a historical reason: the birth of the hippie culture in the 1960s. It seems that true innovation needs rebels, who are ready to challenge the establishment.

But how can we unleash the power of new ideas today? We need to give scientists more freedoms. Instead of asking for proposals, I suggest that interdisciplinary panels should try to determine the smartest junior scientists based on their CVs and presentation of ideas, giving them a safe salary for at least 4 years. For the sake of collaboration and support, the successful candidates would then look for a suitable academic team to realize the ideas they have in mind. The corresponding professor or leader of a research team could either accept or decline the applicants. In case of acceptance, the team of the host would get some overheads. I call this approach the "marriage principle," and I believe it could largely replace the proposal-based funding system, which slows everything down. It would create much better chances for new ideas compared to those established earlier on. For sure, such a bottom-up approach, empowering junior scientists, would be more responsive to emerging challenges than the research and development system we have today.

2.8   Re-inventing Innovation

But there are at least two more things that can help to accelerate innovation dramatically: the way we collaborate and the way we handle intellectual property rights (IPR). I will argue that, to catch up with the pace at which our world is changing, the current IPR protection approach is a great obstacle, and that we need a novel co-creation paradigm. In fact, using information and communication systems, we can re-invent innovation. Currently, people don't like to share their best ideas, because they prefer to be rewarded for them rather than allowing other people to become successful or rich on them. Therefore, it often takes years until an idea is shared with the world through a publication or patent. What, if we instead innovated globally and cooperatively from the very first moment? Say, an idea is born in America, and it is shared through a public portal such as github. Then, experts from Asia could work on these ideas just hours later, and those from Europe would build on their results. In such a way, one can create a new research and development paradigm that never sleeps; one that overcomes the limits of a single team; one that embraces "collective intelligence" by integrating diverse perspectives.[13] Such an approach actually produces considerable synergy effects. As my colleagues Didier Sornette and Thomas Maillart have recently shown: $1 + 1 = 2.5$. Specifically, the collaboration of two people on producing open source software creates outcomes that would otherwise take 2.5 people. Already before, Geoffrey West, Luis Bettencourt, some others and I have found evidence for a similar scaling law for cities[14]: productivity that depends on social interactions grows super-linearly with city size (namely according to a power law with an exponent around 1.2). This is probably the main reason for the dramatic on-going urbanization of the world.

Now, with Internet forums of all kinds, something like virtual cities have grown. Many citizen science projects (and also the famous Polymath project on collaborative mathematics[15]) underline that a crowd-based approach can outperform classical approaches in research and development. So, given the great advantages of collaboration, what are the main obstacles to the immediate sharing of ideas? I would say, mainly the lack of proper incentive systems. Researchers live on two kinds of rewards: their limited salary and the applause they get in terms of citations, i.e. the mentions they receive by fellow scientists. Many of them would not like to share their best ideas with others, before they have published them, nor would they like a

---

[13] A.W. Wooley et al. (2010) Evidence for a collective intelligence factor in the performance of human groups. Science 330, 886–888; S.E. Page (2008) The Difference: How the Power of Diversity Creates Better Groups, Firms, Schools, and Societies (Princeton University); Networked minds: Where human evolution is heading, http://papers.ssrn.com/sol3/papers.cfm?abstract_id=2537497; D. Helbing and S. Klauser, How to make democracy work in the digital age, http://www.huffingtonpost.com/entry/how-to-make-democracy-work-in-the-digital-age_us_57a2f488e4b0456cb7e17e0f

[14] L.M.A. Bettencourt et al. (2007) Growth, innovation, scaling and the pace of life in cities. Proceedings of the National Academy of Sciences of the USA 104, 7301–7306.

[15] T. Gowers and M. Nielsen (2009) Massively collaborative mathematics. Nature 461, 879–881.

company walk away with the idea and make a lot of money on it without sharing the profits in a fair way. Patents and other intellectual property rights are intended to protect the commercial value of ideas and thereby to stimulate innovation. However, in the area of digital products, patents seem to be more an obstacle to innovation rather than a catalyst for it.

Patents on ideas are a bit as if everyone would own a certain number of words and would charge others for using them—this would certainly considerably obstruct the exchange of ideas. In fact, it has recently been difficult to enforce hardware and software patents, and we see ever more patent deals between competing companies. It has also been found that opportunities to watch music videos free of charge can promote their overall sales. Interestingly, Elon Musk leading the electro-car company Tesla Motors has recently decided to allow others to use their patents. Furthermore, Google TensorFlow and OpenAI have made their artificial intelligence algorithms open source. All this might indicate that a paradigm shift regarding Intellectual Property Rights in favor of Open Innovation is just around the corner.

2.9 Micropayments Are Better Than Protecting Intellectual Property

So, why not pursue an entirely different intellectual property rights (IPR) approach—perhaps as replacement, perhaps in parallel to the intellectual property approach of today? At the moment, we are trying to prevent the copying of digital products and handle them as if they were material ones. However, it is the nature of information that it can be reproduced as often as we like and almost for free. Information is a virtually unlimited resource. In contrast to material resources, it would allow us to create benefits and opportunities for all, thereby overcoming poverty and conflict. Turning information into a scarce resource and protecting it from duplication is stupid. It is against the "nature" of information. Therefore, what if we allowed copying, but introduced a micropayment system that ensures that every copy generates some profit for the originator? Under such circumstances, we would love duplication!

Rather than complaining about copies, we should make it easy to be paid for the results of creative and innovative activity. Remember that, some time back, Apple's iTunes made it simple to buy songs and download them, for 99 cents each, thereby overcoming the need for individual negotiations. It would be great to have a similarly simple, automatic compensation scheme for digital products, ideas and innovations. Today's powerful text mining and pattern recognition algorithms could be the basis. Then, whenever another person's or company's idea would be used, there would be an automatic payment, which could depend on the amount of investment made, the novelty of the invention, the advance (the innovativeness), and the "age" of the innovation. This would overcome obstacles like patents, and it would encourage cooperative innovation activities without having to worry that someone could steal an idea.

In perspective, competition would be replaced by co-opetition, i.e. a combination of competition and collaboration. Co-opetition would accelerate innovation enormously. In fact, an Innovation Accelerator should make it simple for people with compatible interests to find together and join their diverse skills in order to carry out an interesting project together. It is rather unrealistic to expect that a scientist who does fundamental research would eventually do applied research, and then develop a product and establish an own spin-off company. This does not happen very often. It would be much better to create alliances between Universities and Universities of Applied Sciences and start-ups or companies that take care of the different steps of this maturation process, ideally all on one innovation campus. I would even say, we need entirely new R&D and education environments for digital innovation, as we have special research and education systems for agriculture, for engineering (the industrial society), and for business (the service society). Such innovation and education environments should make use of the opportunities created by modern platforms such as edx, Khan Academy, Udacity, Mendeley or ResearchGate, but also of Amazon Mechanical Turk, Innocentive and other crowd sourcing, crowd funding and citizen science platforms. To ensure an efficient transfer of knowledge, it is now important to create an innovation ecosystem,[16] where the best of all knowledge and ideas can come together. What's the big idea that you will contribute to it?

Further Reading

Balietti, S., Goldstone, R., Helbing, D.: Peer review and competition in the Art Exhibition Game. Proc. Natl. Acad. Sci. U. S. A. **113**, 8414–8419 (2016)

Helbing, D.: The automation of society is next: How to survive the digital revolution (CreateSpace, 2015)

Helbing, D., Balietti, S.: How to create an innovation accelerator. http://epjst.epj.org/EPJ Spec. Top. **195**, 101–136 (2011). http://epjst.epj.org/index.php?option=com_article& access=standard&Itemid=129&url=/articles/epjst/abs/2011/04/epjst195004/epjst195004.html

Johnson, J., Buckingham Shum, S., Willis, A., Bishop, S., Zamenopoulos, T., Swithenby, S., MacKay, R., Merali, Y., Lorincz, A., Costea, C., Bourgine, P., Louçã, J., Kapenieks, A., Kelley, P., Caird, S., Bromley, J., Deakin Crick, R., Goldspink, C., Collet, P., Carbone, A., Helbing, D.: The FuturICT education accelerator. EPJ Spec. Top. **214**, 215–243 (2012). http://link.springer.com/content/pdf/10.1140%2Fepjst%2Fe2012-01693-0.pdf

Schich, M., Song, C., Ahn, Y.Y., Mirsky, A., Martino, M., Barabási, A.L., Helbing, D.: A network framework of cultural history. Science **345**, 558–562 (2014). http://www.sciencemag.org/content/345/6196/558.short

---

[16]D. Helbing, Distributed collective intelligence: The network of ideas, https://www.edge.org/response-detail/26194

van Harmelen, F., Kampis, G., Börner, K., van den Besselaar, P., Schultes, E., Goble, C., Groth, P., Mons, B., Anderson, S., Decker, S., Hayes, C., Buecheler, T., Helbing, D.: Theoretical and technological building blocks for an innovation accelerator. EPJ Spec. Top. **214**, 183–214 (2012). http://link.springer.com/content/pdf/10.1140%2Fepjst%2Fe2012-01692-1.pdf

Chapter 3
The Hidden Danger of Big Data

Carlo Ratti and Dirk Helbing

With big data, we can multiply our options and filter out things we don't want to see. But there is much to be said for making discoveries through pure serendipity: contingency and randomness often furnish the transformational or counterintuitive ideas that propel humanity forward.

In game theory, the "price of anarchy" describes how individuals acting in their own self-interest within a larger system tend to reduce that larger system's efficiency. It is a ubiquitous phenomenon, one that almost all of us confront, in some form, on a regular basis.

For example, if you are a city planner in charge of traffic management, there are two ways you can address traffic flows in your city. Generally, a centralized, top-down approach—one that comprehends the entire system, identifies choke points, and makes changes to eliminate them—will be more efficient than simply letting individual drivers make their own choices on the road, with the assumption that these choices, in aggregate, will lead to an acceptable outcome. The first approach reduces the cost of anarchy and makes better use of all available information.

This opinion piece by **Carlo Ratti** and Dirk Helbing was first published on August 16, 2016, by Project Syndicate under the URL https://www.project-syndicate.org/commentary/data-optimization-danger-by-carlo-ratti-and-dirk-helbing-2016-08

C. Ratti (✉)
Massachusetts Institute of Technology, Cambridge, MA, USA
e-mail: ratti@mit.edu

D. Helbing (✉)
ETH Zurich, Zürich, Switzerland

TU Delft, Delft, Netherlands

Complexity Science Hub, Vienna, Austria
e-mail: dhelbing@ethz.ch

The world today is awash in data. In 2015, mankind produced as much information as was created in all previous years of human civilization. Every time we send a message, make a call, or complete a transaction, we leave digital traces. We are quickly approaching what Italian writer Italo Calvino presciently called the "memory of the world": a full digital copy of our physical universe.

As the Internet expands into new realms of physical space through the Internet of Things, the price of anarchy will become a crucial metric in our society, and the temptation to eliminate it with the power of big data analytics will grow stronger.

Examples of this abound. Consider the familiar act of buying a book online through Amazon. Amazon has a mountain of information about all of its users—from their profiles to their search histories to the sentences they highlight in e-books—which it uses to predict what they might want to buy next. As in all forms of centralized artificial intelligence, past patterns are used to forecast future ones. Amazon can look at the last ten books you purchased and, with increasing accuracy, suggest what you might want to read next.

But here we should consider what is lost when we reduce the level of anarchy. The most meaningful book you should read after those previous ten is not one that fits neatly into an established pattern, but rather one that surprises or challenges you to look at the world in a different way.

Contrary to the traffic-flow scenario described above, optimized suggestions—which often amount to a self-fulfilling prophecy of your next purchase—might not be the best paradigm for online book browsing. Big data can multiply our options while filtering out things we don't want to see, but there is something to be said for discovering that 11th book through pure serendipity.

What is true of book buying is also true for many other systems that are being digitized, such as our cities and societies. Centralized municipal systems now use algorithms to monitor urban infrastructure, from traffic lights and subway use, to waste disposal and energy delivery. Many mayors worldwide are fascinated by the idea of a central control room, such as Rio de Janeiro's IBM-designed operations center, where city managers can respond to new information in real time.

But with centralized algorithms coming to manage every facet of society, data-driven technocracy is threatening to overwhelm innovation and democracy. This outcome should be avoided at all costs. Decentralized decision-making is crucial for the enrichment of society. Data-driven optimization, conversely, derives solutions from a predetermined paradigm, which, in its current form, often excludes the transformational or counterintuitive ideas that propel humanity forward.

A certain amount of randomness in our lives allows for new ideas or modes of thinking that would otherwise be missed. And, on a macro scale, it is necessary for life itself. If nature had used predictive algorithms that prevented random mutation in the replication of DNA, our planet would probably still be at the stage of a very optimized single-cell organism.

Decentralized decision-making can create synergies between human and machine intelligence through processes of natural and artificial co-evolution. Distributed intelligence might sometimes reduce efficiency in the short term, but it will ultimately lead to a more creative, diverse, and resilient society. The price of anarchy is a price well worth paying if we want to preserve innovation through serendipity.

Chapter 4
Machine Intelligence: Blessing or Curse? It Depends on Us!

Dirk Helbing

Artificial Intelligence (AI) can help us in many ways. Particularly when combined with robotics, AI can make our everyday life more comfortable (e.g. clean our home). It can perform hard, dangerous and boring work for us. It can help us to save lives and cope with disasters more successfully. It can support patients and elderly people. It can support us in our everyday activities, and it can make our lives more interesting. I believe that most of us would like to benefit from these unprecedented opportunities. So far, however, any technology came along with side effects and risks. As I will show, people may lose self-determination and democracy, companies may lose control, and nations may lose their sovereignty, if we do not pay attention. In the following, I describe a worst-case and a best-case scenario to illustrate that our society is at a crossroads. It is crucial now to take the right path.

For a long time, research in Artificial Intelligence (AI) has made frustratingly little progress. However, such progress is exponential. For decades, it is slow and hard to notice. But then, everything happens very quickly and at an accelerating pace. According to Ray Kurzweil, a technology guru in the Silicon Valley working for the Google Brain project, computers will surpass the capacity of the human brain before 2030 and the capacity of all human brains before 2060. Such forecasts have long been considered science fiction. But now there are deep learning algorithms, and AI can learn by itself, making explosive progress.

This article was written by Dirk Helbing for Deutsche Telekom's Digital Responsibility initiative and first published under the URL https://www.telekom.com/en/company/digital-responsibility/details/machine-intelligence--blessing-or-curse--it-depends-on-us--429070

D. Helbing (✉)
ETH Zurich, Zürich, Switzerland

TU Delft, Delft, Netherlands

Complexity Science Hub, Vienna, Austria
e-mail: dhelbing@ethz.ch

Since a number of decades already, computers are better at playing chess. In the meantime, they are better in almost all strategic games people like to play. Now, IBM's Watson computer is able to win game shows—and not only this. Watson is also better in coming up with many medical diagnoses. Moreover, about 70% of all financial trades are performed by autonomous computer algorithms. Soon, we may use self-driving cars that drive better than humans. Algorithms come also increasingly close to human abilities in recognizing handwritings, listening to languages, translating them, and identifying patterns. As 90% of today's jobs are based on these abilities, it will soon be possible to replace all routine jobs by computer algorithms or robots, which will perform better, never get tired, never complain, and will not have to pay social insurance or taxes.

In my contribution to John Brockman's book "What to think about machines that think",[1] I summarize the situation as follows: "The explosive increase in processing power and data, fueled by powerful machine learning algorithms, finally empowers silicium-based intelligence to overtake carbon-based intelligence. Intelligent machines don't need to be programmed anymore, they can learn and evolve by themselves, at a speed much faster than human intelligence progresses."

Jim Spohrer at IBM thinks: While AI will be our tool in the beginning, robots may soon be our teammates, and then our coaches. Therefore, I believe that AI will eventually establish new intelligent species, new forms of "life". However, will highly advanced AI continue to serve humans, will it enslave us, or will it be disinterested in us (as the deep learning expert Jürgen Schmidhuber assumes)? At present, nobody can answer this question. Steve Wozniak, the co-founder of Apple, put it like this[2]: "... I agree that the future is scary and very bad for people. If we build these devices to take care of everything for us, eventually they'll think faster than us and they'll get rid of the slow humans to run companies more efficiently. [But:] Will we be the gods? Will we be the family pets? Or will we be ants that get stepped on? I don't know ..." The technology visionaries Bill Gates and Elon Musk have recently raised similar concerns about superintelligence. What makes them so nervous?

In "What to think about machines that think" I analyze: "Humans weren't very good at accepting that the Earth was not the center of the universe, and they still have difficulties accepting that they are the result of chance and selection, as evolutionary theory teaches us.

Now, we are about to lose the position of the most intelligent species on Earth. Are people ready for this?"

In short: "No, we are not ready for this, but we need to get ready as quickly as we can." Let me start with a worst-case scenario, before I discuss a best-case one.

---

[1] J. Brockman (ed.) What to Think About Machines that Think (Harper Perennial, 2015).

[2] See http://www.computerworld.com/article/2901679/steve-wozniak-on-ai-will-we-be-pets-or-mere-ants-to-besquashed-our-robot-overlords.html, accessed on January 24, 2016.

4.1 A Worst-Case Scenario

In the past, whenever people have raised concerns that AI may take over the world, experts said we could always pull the plug, and this would solve the problem. Unfortunately, this is not true. First of all, many parts of our economy and society would not work anymore, if we turned intelligent machines off. This includes our money and our communication system, increasingly many critical infrastructures, and their protection. Second of all, in some sense, AI has already escaped in the real world. Since Google made their machine learning software "Tensorflow" open source,[3] anyone can use it as he or she likes, including criminals and terrorists. As a consequence of this AI proliferation, we can expect to see a further rise in cybercrime and increasing cyberwar threats. Note that cybercrime already causes economic losses of three trillion dollars annually, and it is growing exponentially.

Nevertheless, my main concern is not that AI might take over the world. It is rather that a few people might try to use AI technology to take over the world. Let me shortly summarize the related developments here: In the 1970s, Chile was experimenting with a third political system besides communism and capitalism: the cybernetic society, inspired by the work of Norbert Wiener (1894–1964). In this system, factories had to regularly report their production numbers to a control center, which told them how to adapt their production. Given the state of information technology at this time, the system worked surprisingly well. In particular, it helped the government to cope with a general strike. However, the CIA supported a military coup in the country, which ended Chile's government on September 11, 1973, with the suicide of its president Salvador Allende.

It seems that, in the meantime, new kinds of cybernetic societies have been built, which use mass surveillance data. Singapore, for example, considers itself a social laboratory.[4] Justified by terror threats, large amounts of personal data are collected about every single citizen. These data are now being fed into AI systems, which learn how every citizen behaves. China's brain project is an example for this. In other words, using our personal data, digital doubles of ourselves are being created, which are thought to replicate our own decisions and behaviors. However, this replication is not perfect, and for such reasons, methods have been developed to manipulate people's decisions and to control their behaviors.

These approaches build on the work of Burrhus Frederic Skinner (1904–1990). He put animals such as rats, pidgins, and dogs into so-called "Skinner boxes" and applied certain stimuli to them. By means of rewards (such as food) and punishments (such as electrical shocks), he managed to condition animals to perform certain kinds of behaviors. Today, we are literally all experimental subjects of companies such as

---

[3]See https://www.tensorflow.org/, accessed on January 24, 2016.
[4]See   http://foreignpolicy.com/2014/07/29/the-social-laboratory/   and   http://www.internationalinnovation.com/predicting-how-people-think-and-behave/, accessed on January 24, 2016.

Google and Facebook,[5] who are performing millions of automated experiments with us every day. Our Skinner box is the "filter bubble"[6] created around us, i.e. the personalized information we receive about the world. In this way, we are exposed to certain kinds of stimuli. AI systems learn how we respond to them and how these stimuli can be used to trigger certain behavioral responses.

In other words, the trend goes from programming computers to programming people.[7] This manipulation is often so subtle that we would not even notice it. What we perceive as our own decisions is now, in fact, often decided by others and secretly imposed on us. This technology, developed to personalize advertisements and make them more effective, has now become a tool of politics, too. "Big nudging",[8] the combination of the "nudging" approach from behavioral economics with "big data" about all of our behaviors, is being used to influence public opinions and election outcomes.

However, the nudging approach is not powerful enough to reach a healthy and environmentally friendly behavior of a nation's population.[9] For such reasons, more effective feedback mechanisms such as personalized prices are being developed. The "citizen score",[10] as it is currently implemented in China, is the first step. Here, everything people do would get plus or minus points: the shopping behavior as well as the links clicked in the Internet. The political opinion is evaluated as well as the behavior of one's social network.

The citizen score will determine credit conditions, the jobs one can get, and whether one is allowed to travel to certain countries or not. In other words, the citizen score is being used to create a modern "caste system". In case of resource shortages, the score would determine who is entitled to get some of the scarce resource and who is not. In other words, the "citizen score" is a mechanism that allows one to play "judgment day" based on arbitrary criteria chosen by "big governments" or "big business".

Most likely, big nudging and citizen score technologies have not only been deployed in Singapore and China. According to "nudging pope" Richard Thaler, no less than 90 countries have established "nudging units" in recent years. So far, little is publicly known about these units. It must be assumed, however, that they involve powerful IT infrastructures, which are fed with personal data collected by

---

[5]See, for example, http://www.wsj.com/articles/furor-erupts-over-facebook-experiment-on-users-1404085840 and http://www.pnas.org/content/111/24/8788.full, accessed on January 24, 2016.

[6]E. Pariser, Filter Bubble: What the Internet Is Hiding from You (Penguin, 2011).

[7]See https://www.youtube.com/watch?v=KlWeuK46_nA and https://www.youtube.com/watch?v=pplhyw-vEWg, accessed on January 24, 2016.

[8]http://www.spektrum.de/news/big-nudging-zur-problemloesung-wenig-geeignet/1375930, accessed on January 24, 2016.

[9]http://www.theguardian.com/commentisfree/2011/jul/19/nudge-is-not-enough-behaviour-change, accessed on January 24, 2016.

[10]http://www.computerworld.com/article/2990203/security/aclu-orwellian-citizen-score-chinas-credit-score-system-isa-warning-for-americans.html, accessed on January 24, 2016.

mass surveillance and profiling activities of private companies. Such infrastructures are used to run autocratic countries such as Saudi Arabia.

The likely goal of big nudging and citizen scores is to control a society similar to the Singaporean or Chinese model. The fundamental idea is that of a data-driven cybernetic society, which is controlled by a "benevolent dictator". Such an approach is certainly not compatible with democratic principles and constitutional rights. Should the Singaporean or Chinese model be applied in the aforementioned 90 states, democracies worldwide might be in great danger.[11]

The problem is that a digital power grab is easily possible and irreversible. For example, whoever has access to a big nudging infrastructure may be able to determine the result of an election.[12] Furthermore, terror attacks or other events that traumatize the public may be used to restrict democratic principles. This happened in France, where mass surveillance is already being used as an instrument to keep the own citizens in check and to suppress the opposition, as criticized by a high-level UN committee.[13] The example of Poland shows as well, how easy it is to demolish democratic institutions such as the constitutional court and the freedom of the press. We can see similar developments in Hungary, where the constitution is about to be changed,[14] and Turkey. In Turkey, both the opposition and the Kurdish minority are already being suppressed.

Given the incredible potential to inflict harm and violate basic human rights created by the confluence of technologies as discussed above, we urgently need initiatives to implement the following measures as quickly as possible:

- The above instruments should be democratically controlled by the parliament and not an exclusive tool of the chancellor or president, the government, the military, or secret service.
- It also makes sense to give opposition parties access to such information systems in order to ensure a reasonable balance of power. (Remember that complex systems such as our society require pluralistic perspectives to be well understood and controlled.)
- The use of these tools should be based on democratic mandate and scientific principles. They should be operated by interdisciplinary teams of leading scientists (including psychologists, sociologists, economists, computer scientists, and

---

[11]Moreover, as the Stanford Prison Experiment has shown, any system that creates too much difference in power between those who decide and those who have to obey will sooner or later turn bad and get out of control.

[12]http://www.pnas.org/content/112/33/E4512.abstract, accessed on January 25, 2016.

[13]https://netzpolitik.org/2016/un-sonderberichterstatter-kritisieren-frankreichs-flaechendeckende-ueberwachung/ and http://www.ohchr.org/EN/NewsEvents/Pages/DisplayNews.aspx?NewsID=16966&LangID=E, accessed on January 25, 2016.

[14]http://www.focus.de/politik/ausland/kontaktverbot-und-umsiedlung-terror-notstand-geheimer-notfallplan-koennteorban-zu-ungeahnter-macht-verhelfen_id_5234034.html, accessed on January 25, 2016.

complexity scientists). These groups need to be open to international exchange and report about their activities at public international conferences.

- Ethical oversight should also be ensured.
- Personal data should be anonymized and breaches of privacy punished.
- Transparency about on-going activities would be important. It must be recorded who uses the system in what way, and the uses need to be regularly reported to the public in comprehensive documents.
- Opt-out (at least from scoring and big nudging) should be offered to ensure informational self-determination. (Note that this will also promote trustworthy uses of these methods.)
- If social experiments have caused undesirable side effects, victims should be properly compensated.

Secret services would probably want to have separate access to these information systems, but some principles should nevertheless apply:

- The use of these tools should be recorded. Large-scale nudging should be forbidden. Individual-level nudging in a limited number of cases may be acceptable.
- Mass surveillance should be also forbidden. Deanonymization should be limited to a small number of people and democratically controlled.

Private companies would have to follow the new European General Data Protection Regulation, and the government would have to enforce compliance not only of big IT companies, but also of the often relatively unknown companies trading with our personal data.

Democracy certainly deserves a "digital upgrade",[15] but it would be a disaster for the future of our planet, if democracy would be extinct, i.e. if we didn't have a competition between different political systems anymore. It has been found that, in the long term, only democracies can live peacefully with each other, thanks to the effectiveness of their principle of balancing different interests (through subsidiarity, federal organization, separation of powers, and citizen participation). It should be remembered that these institutional designs as well as human rights and our justice system are the lessons learnt over hundreds of years, including two World Wars.

Why am I certain that a democratic, data-driven approach will be superior? Because we have had a similar historical case: the competition between centralized communist regimes which were controlled from the top down, with federally organized capitalist systems which were controlled to a greater extent from the bottom up. Capitalism won because innovation mainly happens from the bottom up. (Most of the richest people on Earth have made their money within just a few decades with entirely new business models. Some of these businesses were started in garages by students who did not finish their degree.)

---

[15]D. Helbing and E. Pournaras, Build Digital Democracy, Nature 527, 33–34 (2015): http://www.nature.com/news/society-build-digital-democracy-1.18690

In fact, comprehensive analyses of empirical data show that the transition from autocracy to democracy can yield a boost in economic growth. A transition in the opposite direction, however, leads to a loss of sociopolitical capital in the medium term and a loss of economic growth in the long term.[16] Thus, the price of losing democratic principles is high.

I am also involved in a scientific project analyzing virtual (gaming) worlds. Here, we find as well that the worlds using automatic penalty mechanisms similar to the citizen score are not only less attractive, but also less innovative.

I certainly don't want to criticize Singapore or China, here. It is possible that their governance models are the best solutions for their societies in their current historical situation. However, I seriously doubt that this governance approach would be suited as models for the rest of the world. They should not be copied by democratic states. We need to elaborate a different model for them (see the best case scenario discussed below).

Note that Singapore's success does not only rest on its data-driven approach. It was also a tax heaven, and it imports innovations from the US, Germany, Switzerland etc. to a much larger extent than other countries. Without these imports, it would be much weaker in terms of innovation. I am certainly not questioning the import of innovations from elsewhere, but I want to point out that it takes more liberal settings in other countries to produce these innovations in the first place. Singapore knows this,[17] and that's why the country is now trying to grow spaces that allow for some degree of "creative disorder".

Many times, China has also been proposed as a political model to copy. However, despite China's impressive development, the average living standard is still lower than in many democratic countries. Moreover, it is currently faced with acute environmental problems and huge market turbulences. So, it increasingly feels the limitations of centralized control, and it knows that it needs to become more pluralistic and democratic to enable further development.

Finally, it is noteworthy that none of the IT superpowers such as the USA, China, or Singapore has a city in the top ten list of most livable cities in the world. Thus, how can we expect that these governance models would lead to societies with the highest quality of life? If companies like Google could create paradise on Earth, why then is San Francisco not the most attractive city on the planet? Instead, the most attractive cities are all located in countries, which make sure to balance the interests of all stakeholders, including civil society.

In conclusion, the self-determination of people is currently at stake, which is a big concern. Big nudging, citizen scores, and implants could lead to digitally enabled slavery. However, this is not only endangering the freedom of people. It is also endangering the sovereignty of companies and entire countries. This is not just due to mass surveillance and espionage. AI systems can be used to spot the weaknesses of IT systems and people, by sending certain stimuli and recording the responses. In

---

[16]H.H. Nax and A.B. Schorr, Democracy-growth dynamics for richer and poorer countries, http://papers.ssrn.com/sol3/papers.cfm?abstract_id=2698287, accessed on January 24, 2016.

[17]http://www.zeit.de/2015/25/singapur-image-innovation-unterwelt, accessed January 24, 2016.

this way, it is not only possible to learn how to manipulate people (as discussed above). It is also possible to learn how to control IT systems and critical infrastructures. In fact, even autonomous AI systems can be externally steered: since they respond to their information inputs, this can be used to manipulate their outputs. As a consequence, whoever has the most powerful AI system might be able to control all other AI systems and, in this way, all the companies, institutions and people manipulated by them. The explosive evolution of technologies such as quantum computing, memristor technologies, and light-based LiFi communication, implies a race for global dominance.

In other words, technologies intensify the race to control the world and its resources. Today, 62 people are said to control as much capital as 50% of people on this planet.[18] The following people lead the ranking: Bill Gates (Microsoft, USA), Amancio Ortega (fashion, Spain), Warren Buffet (finance, USA), Jeff Bezos (Amazon, USA), Carlos Slim Helu (telecommunication, Mexico), Larry Ellison (Oracle, USA), Mark Zuckerberg (Facebook, USA), Charles and David Koch (oil and various products, USA), Liliane Bettencourt (L'Oreal, France), Michael Bloomberg (finance data, USA), Larry Page (Google, USA), Sergey Brin (Google, USA). We see that business with data, software and information and communication technologies is outpacing most classical business models, and I expect that we might see a further rapid concentration process, until the world is controlled by very few people.

In fact, while I am generally not opposed to free trade agreements, it is to be expected that the impending TTIP and TISA agreements will accelerate this concentration process. In the end, most money, power and resources would end up in the hands of very few people (most likely not in Europe). These people could decide the fate of the planet like dictators. Would these people pay a basic income or engage in other solutions that would allow the many unemployed people to survive, whose jobs will be taken by robots and artificial intelligence? Or would we face a global war, until we remain with, say, a billion highly qualified people or so, needed to run a data-driven brave new world? Several IT companies have started to build their own rockets. Their ambition to control the universe is hard to ignore.

However, I don't think that any kind of global dominance would be good for our planet and for humanity, and I am not alone. More than 20,000 people have recently signed a petition that we should not use AI as a weapon against humans.[19] I believe we should rather engage in a cooperative AI paradigm and distance ourselves from "big nudging", "citizen scores" and other approaches, which may be used to control millions, perhaps billions of people in a centralized and top-down way, which could easily end in the most totalitarian regime ever.

---

[18]http://m.spiegel.de/wirtschaft/a-1072576.html accessed on January 24, 2016 and http://www.forbes.com/forbes/welcome/#version:realtime on January 18, 2016.

[19]http://futureoflife.org/open-letter-autonomous-weapons/ accessed on January 24, 2016.

4.2   A Best-Case Scenario

Fortunately, there are positive perspectives, too. We are just about to step into a new era of history—the digital society and economy 4.0. If we want this transformation to succeed, it is important that we create opportunities for everyone: business, politics, science, and citizens alike. With new information and communication technologies, this can now be accomplished more easily than ever. The good news is that the digital economy is not a "zero sum game". It allows us to overcome the exclusive competition, which we have had in the material world and the old economy. Now, competition becomes compatible with cooperation. Therefore, "co-opetition" will be the new paradigm, and if we manage to create a suitable legal framework, everyone could have a prosperous life.

The benefits of open information exchange are becoming increasingly evident. More and more people understand that sharing information often increases the value of information, inventions, and companies. If properly organized, the digital economy provides almost unlimited possibilities because intangible goods can be reproduced as often as we like and used in zillions of different ways. For example, more and more money will be earned in virtual worlds. This relates not just to computer games; Bitcoin has even shown that bits can be transformed into gold. Almost nobody believed that this were possible.

In fact, if we want to master the challenges humanity is faced with, our economy and society will (have to) be organized in entirely new ways. Without any doubt, in the next three decades to come, the world will see disruptive times, characterized by problems such as the digital revolution, financial and economic crises, climate change (with extreme weather and loss of biodiversity), the energy transformation, demographic challenges (such as ageing and migration), and unstable peace.

I do not think the use of Big Data and Artificial Intelligence will be the one and only solution to the above problems (see the Appendix). Due to the complexity of the world, Big-Data-driven "crystal balls" to predict the future will often fail; predictive analytics and control may be even counter-productive. They may keep us from exploring innovative paths, because AI systems, which are based on backward data, tend to repeat solutions of the past (in extreme cases, this may also be war).

Instead we need a resilient design and operation of our society. This requires diverse system components, modular system designs, and decentralized control paradigms, i.e. bottom-up participation. Such an approach would allow for a flexible adaptation to unexpected events. We also need to increase innovation rates dramatically. Therefore, I expect the new organizational principles of our future society to be collective intelligence and co-evolution in a highly diverse networked econom—an emerging participatory society, which may be viewed as an "innovation ecology".

If we want to boost innovation, the application of big nudging and citizen scores is quite counterproductive. It would promote opportunism and conformism, rather than increasing people's readiness to take risks and to question existing solutions—something, which is absolutely necessary now.

We further need a fundamentally new approach to innovation that puts more emphasis on open innovation than today, in order to offer all of the products and services that are currently not provided by large companies. Citizen science, so-called Fablabs (public centers for communities of digital hobbyists), as well as initiatives to mobilize civil society are becoming increasingly important. The key word is co-creation, which means that citizens can augment information, knowledge, services and products in a largely open information and innovation ecosystem. Obviously, this does not preclude commercialization. On the contrary, it would create opportunities for everyone to earn money with data. The citizens and the customers would become partners. The participatory society of the future will not only build on large global corporations. Businesses of all types and sizes and self-employment will play an even bigger role than today. This is a good thing because monopolies are known to be comparatively little innovative, and they rarely care about products and services that will not generate a significant return, say, of 20%. Nonetheless, big business could also benefit. A rich information ecosystem is like a rain forest, in which many trees are much bigger than the few trees growing in the desert.

In this connection, the OpenAI initiative, which was recently started with a donation of one billion dollars, is quite remarkable. Initiator Elon Musk formulated the goals as follows: "AI should be an extension of individual human wills and, in the spirit of liberty, as broadly and evenly distributed as possible." In addition, however, we need to engage in a responsible innovation paradigm, oriented at creating value-sensitive designs that fit the respective context and culture. In particular, we must design and teach AI systems to act morally and socially. This will change the paradigm of "human-machine interaction" to the paradigm of "human-machine symbiosis". In John Brockman's book "How to think about machines that think" I conclude:

> [On the long run,] Intelligent machines would probably learn that it is good to network and cooperate, to decide in other-regarding ways, and to pay attention to systemic outcomes. They would soon learn that diversity is important for innovation, systemic resilience, and collective intelligence. Humans would become nodes in a global network of intelligences and a huge ecosystem of ideas.

I also believe we need to generate knowledge in real time (as much as this can be done) and share our reflections, judgments and insights more adequately, faster, and worldwide. In the digital age, we must reinvent innovation, from research to publication to teaching, and this requires a new framework that I like to call "Plurality University".

We further need to think more about ways to foster the spirit of experimentation. Too many inventions are merely modest improvements of existing ideas, so-called linear innovation, which extend the life cycle of "old" products. Instead, we need to encourage radically new ideas, sometimes referred to as "disruptive innovations".

The question is, how we can ensure that such innovations will lead to sustainable products that do not harm our society and environment (given that the hope our planet would recover from all stresses and strains by itself has not materialized). For this, we need to measure and price externalities, which refer to the external costs or benefits associated with products, services and interactions. Interestingly, Big Data and the Internet of Things (IoT) make it increasingly possible to do this.

Note that the measurement and pricing of externalities would require less regulation, so it could help to swipe today's over-regulation out of the way. If we were to trade externalities like financial derivatives, this would create entirely new financial markets. That would unleash enormous economic potential. A multi-dimensional financial system would also allow enirely new applications such as self-organizing socio-economic systems, which require various incentive mechanisms. In many cases, the application of decentralization approaches and self-organization principles could increase the resource efficiency by 30–40%.

Therefore, it makes a lot of sense to empower citizens by means of information and communication technologies in order to allow them to make better decisions and contribute more to business and society, and to their digital transformation. If set up well, enabling users, customers, and citizens will lead to better services, better products, better businesses, better neighborhoods, smarter cities, and smarter societies.

For example, digital assistants can help people to behave in a healthier and environmentally friendly way. A GPS-based route guidance system may serve to illustrate this. There, the user can specify the goal, and the digital assistant offers various alternatives to choose from, pointing out the advantages and disadvantages of each. After that, the digital assistant supports the user as good as it can in reaching the goal and in making better decisions. To stimulate people to do more sports and eat more healthy food, it is not needed that the state or a health insurance records everyone's personal information. One can also think of a social media platform that allows people to form their own "health circles". The competition between friends would be able to stimulate healthier behavior. To provide incentives without violating privacy, the state or a health insurance might reward health circles rather than individuals, but this is perhaps not even necessary.

I believe that modern information technology can also help us to reduce conflict in the world, namely by mitigating the competition for scarce resources. This can be achieved by the combination of several measures. First, resources need to be used more efficiently, as discussed before. Second, recycling techniques could be considerably advanced. Third, the principles of the sharing economy could be applied to an increasing number of areas of social and economic life, including how urban space is managed and used. This would enable a higher standard of living for more people while decreasing the consumption of resources. In order to reduce war and terrorism, we certainly need to pay more attention to the living conditions in the rest of the world.

Furthermore, we must learn that, in a multicultural society, punishment mechanisms often do not cause social order, but rather escalation of conflict. This has been observed not only in the Middle East, but also in Ferguson, and many other places. Therefore, we need new mechanisms to promote coordination and cooperation in a multi-cultural world. Suitable reputation mechanisms are promising in this regard, but also qualification, competition, communication and matching mechanisms.

Last but not least, engaging in a "Cultural Genome Project" could achieve a better understanding of the success principles, on which different cultures are built. This would allow us to combine them in innovative ways and enable us to generate new social and economic value. The greatest potential of this approach lies directly on today's cultural fault lines. Some of these cultural success mechanisms will, for example, be built into the Nervousnet platform,[20] so that its "data for all" approach will lead to responsible use.

Nervousnet (see nervousnet.info) is an open and participatory, citizen-run Internet-of-Things platform that will support (1) real-time measurements of the world around us, (2) its scientific understanding, (3) awareness of the implications of various decision alternatives, (4) real-time feedback to support self-organization, and (5) collective intelligence. The project takes informational self-determination seriously. Its data storage is decentralized and various procedure are used to anonymize, encrypt and "forget" data. Users can decide for themselves which kind of data they want to produce for themselves or share with others.

In addition, imagine that all the data you generate is sent to a personal data store, where it can be sorted and managed by category. Given appropriate technical solutions and legal regulations, you would then be able decide what kind of data to share with whom, and for what purpose. Thus, more trusted companies would have access to more data. This would stimulate competition for trust, and the data-driven society would be built on trust again.

It is high time that we start to work on this best case scenario. First of all, we are forced to be much more innovative than today. Second, it would bring great benefits to everyone to implement these proposals. It seems that the USA have already started to invest in a new strategy. They are betting on reindustrialization on the one hand, and citizen science and combinatorial innovation on the other. Even Google has embarked on a new strategy with the founding of Alphabet, which aims to make the company less dependent on personalized advertising. And Apple has recognized the value of privacy as a competitive advantage.

Finally, people increasingly understand that the digital economy is not a zero-sum game. In the area of the Internet of Things, Google has engaged in open innovation. Tesla Motors has opened up many of its patents, and many billionaires have recently promised to donate large sums of money for good. So, we see many signs of change. The only question is when Europe will finally make use of the fantastic opportunities offered by the digital revolution. We are entering a digital age that increasingly frees itself of material limitations. This is absolutely fascinating!

---

[20]D. Helbing and E. Pournaras, Build Digital Democracy, Nature 527, 33–34 (2015): http://www.nature.com/news/society-build-digital-democracy-1.18690http://futurict.blogspot.ch/2014/09/creating-making-planetary-nervous.htmlhttp://futurict.blogspot.ch/2015/08/smart-data-running-internet-of-things.htmlhttp://futurict.blogspot.ch/2016/01/nervousnet-towards-open-and.html　　　all accessed January 24, 2016.

Appendix: Some Common Pitfalls of Data-Driven Technologies

In the past couple of years, the concept of Big-Data-driven and Artificial-Intelligence-based Smart Nations has spread around the globe. Without any doubt, these technologies offer interesting potentials to improve political decision-making and the state of the world. However, there are also a number of issues that need to be considered[21]:

Big Data Analytics

- In classification problems, errors of first and second kind will occur, which implies unfairness, if decisions cannot be challenged and corrected. Current algorithms to identify terrorists are actually quite bad. They produce too long lists of suspects, and "one does not anymore see the trees for the forest."
- Using more data is not necessarily better: it may lead to over-fitting. In large datasets there are always some patterns and correlations by coincidence. In many cases, however, these patterns are meaningless, or they don't imply causality. This might lead to wrong conclusions, if statistical significance and causality are not ensured (which is often the case today).
- Some data-driven findings may lead to decisions that discriminate people and, thereby, violate constitution or law. Suppose we let people pay different health insurance rates dependent on what they eat. Then, we will for sure end up with different rates for women and men, for Christians, Muslims, and Jews. Such implicit discrimination is to be avoided, but common Big Data methods don't take care of such issues.

Artificial Intelligence (AI)

Such systems can handle huge amounts of information, but:

- errors may still occur due to relevance, inconsistency or incompleteness of information, ambiguity, context-dependence, etc.
- the goal function may be specified in an improper way, and by modification of the goal function one will often get completely different results as a consequence of

---

[21]For a detailed discussion see: D. Helbing, Societal, Economic, Ethical and Legal Challenges of the Digital Revolution: From Big Data to Deep Learning, Artificial Intelligence, and Manipulative Technologies, Jusletter IT (2015), see http://papers.ssrn.com/soL3/papers.cfm?abstract_id=2594352; D. Helbing, B.S. Frey, G. Gigerenzer, E. Hafen, M. Hagner, Y. Hofstetter, J. van den Hoven, R.V. Zicari and A. Zwitter, Digitale Demokratie statt Datendiktatur, Spektrum der Wissenschaft 1/2016, see http://www.spektrum.de/pdf/digital-manifest/1376682

"parameter sensitivity"; this makes results subjective, i.e. dependent on the person who controls the AI system.
- if AI systems are not programmed as tools, but able to learn and evolve, they may start to take unpredictable decisions and behave maliciously.
- if people are involved in defining the training data, they may intentionally or unintentionally introduce biases that are not accounted for, as we currently lack suitable institutional checks and balances regarding such training; if people are not directly involved in selecting the training data, then machine intelligence may run into similar problems as we know them from children that have not received proper moral education or coaching by adults.

Big Nudging

"Big nudging" uses Big Data of a population, AI, and methods from behavioral economics (such as "nudging") to manipulate people in their decision-making and behaviors.

- These systems can be used to let people make stupid mistakes (e.g. spend their money on things they don't need, undermine the security of IT systems, etc.).
- They can be used to manipulate public opinion and democratic elections by means of an almost unnoticeable kind of propaganda and censorship, employing principles from attention economics.
- They amplify the power of those who are allowed to use the system to an extent that is hardly controllable. For example, they can be used for a digital power grab, i.e. to establish and/or stabilize autocratic regimes. These can exploit the data asymmetry to weaken the rule of law or democratic arrangements.

Altogether, the problem of the above three approaches is that their validity is over-rated. They give very few people an extreme amount of power, but are very hard to control. In principle, they can be misused as a "weapon" against the own population. Using "big methods" implies the likelihood of making big mistakes. It's just a matter of time until they will happen.

Further Reading

Helbing, D.: Responding to complexity in socio-economic systems: How to build a smart and resilient society? see http://papers.ssrn.com/Sol3/papers.cfm?abstract_id=2583391

Helbing, D.: The Automation of Society is Next: How to Survive the Digital Revolution. CreateSpace, North Charleston (2015)

Helbing, D.: Societal, economic, ethical and legal challenges of the digital revolution: From big data to deep learning, artificial intelligence, and manipulative technologies, jusletter IT (2015). see http://papers.ssrn.com/soL3/papers.cfm?abstract_id=2594352

Helbing, D. et al.: Eine Strategie für das digitale Zeitalter, Spektrum der Wissenschaft 1/2015. http://www.spektrum.de/news/eine-strategie-fuer-das-digitale-zeitalter/1376083

Spiekermann, S.: Ethical IT innovation: A value-based system design approach. Auerbach (2015)
van den Hoven, J. et al. (eds.): Responsible Innovation 1: Innovative Solutions for Global Issues. Springer, Dordrecht (2014)
van den Hoven, J. et al. (eds.): Handbook of Ethics, Values and Technological Design. Springer, Dordrecht (2015)
More references at http://www.spektrum.de/news/wie-algorithmen-und-big-data-unsere-zukuntbestimmen/1375933

Chapter 5
An Extension of Asimov's Robotics Laws

Jan Nagler, Jeroen van den Hoven, and Dirk Helbing

In a world, where Artificial Intelligence systems will decide about increasingly many issues, including life and death, how should autonomous systems faced with ethical dilemmas decide, and what is required from humans?

In the near future, autonomous vehicles are expected to substantially improve traffic flow and drastically reduce accidents (van Arem et al. 2006). According to the Department of Energy (DOE), self-driving cars could reduce energy consumption in transportation by as much as 90% (Bullis 2011). A reduction of accidents by 90% seems possible as well, which would translate to saving millions of lives every year world-wide (Gao et al. 2014). However, many pressing questions remain regarding how to engineer autonomous cars and, generally, design artificially intelligent systems for safety and other moral values (Deng 2015). What public policies and what regulations would be needed?

Should we design autonomous decision-making solely on the basis of moral values and moral principles, or should self-interests or company policies and market

This article by **Jan Nagler, Jeroen van den Hoven,** and Dirk Helbing was first published on August 22, 2017, as preprint under the URL https://www.researchgate.net/publication/319205931_An_Extension_of_Asimov%27s_Robotics_Laws

J. Nagler (✉)
ETH Zurich, Zürich, Switzerland
e-mail: jnagler@ethz.ch

J. van den Hoven
TU Delft, Delft, Netherlands
e-mail: m.j.vandenhoven@tudelft.nl

D. Helbing (✉)
ETH Zurich, Zürich, Switzerland

TU Delft, Delft, Netherlands

Complexity Science Hub, Vienna, Austria
e-mail: dhelbing@ethz.ch

© Springer International Publishing AG, part of Springer Nature 2019
D. Helbing (ed.), *Towards Digital Enlightenment*,
https://doi.org/10.1007/978-3-319-90869-4_5

forces predominate? How can one design for moral values and develop an ethically aligned design of autonomous systems (IEEE; Ethically Aligned Design Initiative)? A number of proposals for codes and principles have been made (Asilomar AI Principles). We suggest that rather than a grocery list of values we need some clarity in the form of a limited set of basic moral principles dealing with autonomous vehicles.

Early on, the well-known science-fiction writer Isaac Asimov proposed "Three Laws of Robotics", which he augmented by a fourth law later on. These are: *(i) A robot may not injure a human being or, through inaction, allow a human being to come to harm. (ii) A robot must obey the orders given to it by human beings except where such orders would conflict with the First Law. (iii) A robot must protect its own existence as long as such protection does not conflict with the First or Second Law. (iv) A robot may not harm humanity, or, by inaction, allow humanity to come to harm.*

These laws, however, may induce ethical dilemmas in certain critical situations. One of these, the "trolley problem", received a lot of attention in the autonomous driving literature, recently (Bonnefon et al. 2016). Imagine a runaway railway trolley that is about to kill five people working on the trolley's track. Furthermore, assume that you (or the robot) can save them only if a lever is pulled that diverts the trolley onto another track, where it will kill an innocent bystander with certainty. What should you do?

Analogously, if an autonomous car faces a situation where fatalities are inevitable, should it run over a crowd of people on the street, swerve into a smaller group of pedestrians on the walkway or sacrifice the lives of all car passengers by ramming into a concrete wall? Recently Bonnefon et al. studied the preference of test subjects for a number of similar trolley problems in autonomous driving (Bonnefon et al. 2016). They found that even though participants approve of autonomous vehicles that might sacrifice passengers to save others, respondents would prefer not to ride in such vehicles. Respondents would also not approve regulations mandating self-sacrifice. Such regulations would make them less willing to buy or use an autonomous vehicle. Bonnefon et al. concluded that regulating for utilitarian algorithms (that minimize total harm) may paradoxically lead to more casualties by postponing the adoption of safer (autonomous vehicle) technology.

This may sound like an invitation to leave it to car manufacturers to determine the properties of their autonomous vehicles based on market criteria. However, this might result in different safety features. Luxury cars could offer a higher degree of self-protection as compared to cars in the lower price segment, in order to create incentives to buy a more expensive car. However, luxury cars cause more accidents, i.e. they already impose higher risks on others (The Telegraph 2015). Increasing the relative risk imposed on others by price-sensitive safety technology would be highly questionable from an ethical point of view.

So neither a laissez faire market solution, nor a government's advertising utilitarian policy would work. Which policy is fair, given the fact that safety dilemmas of the trolley type are likely to occur, even if infrequent? Responsibility demands to prevent the occurrence of ethical dilemma situations in the first place, and if this is impossible, to take measures to reduce them as much as possible (Van den Hoven et al. 2012). This suggests the introduction of a fifth law of robotics as follows: *(v) Humanity and robots must do everything possible to reduce the occurrence of ethical dilemma situations.* Sometimes, the fifth law may only require better safety features of autonomous vehicles or stricter speed limits. Generally, however, a re-design of systems—encompassing technology, infrastructure and organiza-tion—may have to be considered.

Yet, ethical dilemma situations may sometimes still occur. How should an autonomous vehicle or, more generally, an Artificial Intelligence (AI) system then decide? With better sensor and video technologies and powerful information sys-tems, can we now take better decisions from the perspective of society? Such systems could identify objects and even individual people. Therefore, AI could distinguish between one person or many, a child and an elderly person, an average person and a famous politician, a white person and a person of colour, a person with a job or without, a rich and a poor person, a convicted criminal and a saint, a healthy person and one who may die soon, a person with health or life insurance and without? So, who should die, if an algorithm had to decide? Should people with higher status or life expectancy be protected, because they may contribute more to society?

Some may find it plausible and acceptable to value people and weigh their lives differently. In fact, not all organisations, leaders, policymakers, stakeholders and nations agree on the same fundamental moral principles to build a society on, or to prevent moral hazards and exploitation (Helbing 2013). In particular, age and health are sometimes used as factors to determine medical treatments and health risks. Explicit discussions have come from the EPA (U.S. Environmental Protection Agency) in its 2003 discussion of hazardous air pollutants: *"There is general agreement that the value to an individual of a reduction in mortality risk can vary based on several factors, including the age of the individual, (...) and the health status of the individual."* (Posner and Sunstein 2005).

Recall, however, that Kant, the father of Enlightenment, who inspired modern democratic constitutions, wrote his masterpieces at old age. Van Gogh had a very low social status during his lifetime. Mozart died poor. Beethoven was almost deaf when he wrote his 9th symphony. Degas and Toulouse-Lautrec were handicapped and Monet had impaired sight, but they became three of the most important painters of Impressionism. These individuals have created some of the greatest cultural achievements in the history of humankind. This invariably shows that, even from a broadly utilitarian point of view, health, age, or social status are not suitable criteria to decide who should come to harm.

Consider also that each of us will get old or may fall ill, and that in 99.9% of cases somebody will be around with higher status than yourself. Should rich people be able to buy themselves a higher chance of survival? Should the person who pays more be saved and others sacrificed? This sounds like a profitable business model, but it would substantially harm our society, which—according to the United Nations' Universal Declaration of Human Rights—is built on equality.

In fact, the current rulings of many constitutional courts and ethical committees largely agree that people should *not* be valued differently, considering, for example, status, age, or health, but share a common humanity and human dignity. This is also a lesson learned from fascism and the Holocaust.

Today, robocops are being tested, drones are being used to kill dissidents, and a number of autonomous weapons are in the making (Russell et al. 2015; Future of Life Institute 2015). Even worse, some experts think about AI-based euthanasia (Hamburg 2005). This includes computer-controlled implants that release drugs (or an overdose) to our bodies. Such devices could also be hacked (Hackers remotely kill jeep highway 2015; Hackers reveal nasty new car attacks 2013). Cybersecurity is an arms race, not a problem that can be solved once and forever. Thus, the ethics of autonomous systems may soon affect all of our lives every day.

Furthermore, people with higher education may have a better chance to live long in prospering modern societies. In turbulent times, however, as we will encounter them in an unsustainable world that is increasingly faced with problems such as mass migration, war and terrorism, other skills may be important to survive. Therefore, whatever measure is taken to distinguish the value of people, there are always examples that show the inappropriateness of such a measure, also on purely utilitarian grounds. In particular, it is not well justified to attribute a different value to different individuals, such as the Citizen Score concept does (Storm 2015).

Therefore, we formulate a sixth law of robotics: *(vi) If the application of rules (i)–(iv) leads to ethical dilemmas, which could not be avoided as required by rule (v), decisions should be randomized, giving each person the same weight.* The sixth law is compatible with the equality principle of many constitutions, the idea of human dignity and also with medical ethical imperatives. Namely, bioethics requires that a doctor's decision which patient to treat, in case two wounded appear equally threatened, should be random as seen by an observer—not based on sex, skin colour, wealth, or other features. A society with ubiquitous AI requires a social contract that is impartial, as proposed by the Harvard political philosopher John Rawls with his concept of "the veil of ignorance" (Leben 2017). This implies that, in deciding about the normative principles of a society, one should ignore properties that serve self-interest.

In the following, we will substantiate the sixth robotic law further. Autonomous driving is challenged by ambiguity, uncertainty and complexity arising from incomplete information, imperfect classification, or just from unclear sensory input such as poor sight (Deng 2015). The number of pedestrians on a road might be uncertain, the number of casualties may not be accurately estimated, or people may rather be injured than killed (see Fig. 5.1). However, it appears unrealistic to specify appropriate deterministic decision rules for all probabilistic trolley problems, or just the majority of them.

Assume two cars that are about to collide on a bridge with deadly consequences for both. If one car decided to swerve and fall off the bridge, this would save the passengers of the other car. This, however, would not happen if both cars were in a "minimum self-harm" mode (in which passenger safety is the ultimate priority). Then, both cars would frontally crash into each other and kill all passengers. This

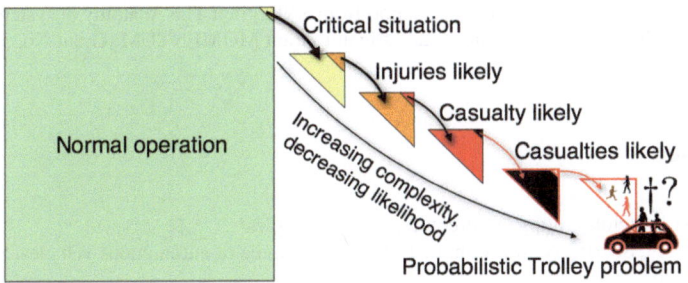

Fig. 5.1 Complexity of the probabilistic trolley problem

dilemma is known in game theory as "chicken game", where paradoxically a "maximum harm" outcome is the expected to occur. A probabilistic decision, in contrast, would guarantee a fair chance of survival.

Next, let us discuss the example of manipulation or hacking. For example, someone may jump on the street and force the vehicle to crash into a concrete wall, or someone may trigger or hack the vehicle sensors to trick it into a dangerous manoeuvre that might put passengers' lives at risk. In such cases, a probabilistic decision rule would make it less likely that an autonomous system could be successfully instrumentalized to harm people.

Finally, let us assume that a Citizen Score, attributing a certain value to everyone's life, would be used to determine who has to die and who will survive, when resources are scarce. Then, this would create a serious moral hazard. "The elite", i.e. the people with the highest scores, would always have the lowest risks and the greatest opportunities. Therefore, why should they make the greatest effort possible to improve the opportunities and risks of all the others, if such an effort will not improve their own situation?

A Citizen-Score-based system would, hence, reduce the chance that the fifth law will be taken seriously and sufficient efforts will be made to avoid ethical dilemmas. A fair probabilistic decision rule, in contrast, would put everyone at the same level of risk, and hence everybody would have an incentive to reduce the number of ethical dilemmatic situations as much as possible.

To conclude, the fifth law demands artificial and human intelligence as well as creativity to avoid ethical dilemmas and critical situations. A successful implementation of the fifth law may require us to change the monetary, financial, and economic system, or even the organization of society altogether (Helbing 2013; Helbing and Pournaras 2015). Today's algorithms with the objective to minimize harm are deterministic (Rasmussen 2012) and, thus, may lead to unnecessary harm (as the above bridge problem has illustrated). Our extension of Asimov's Robotics Laws considers the limitations of optimization, Citizen Scores and utility maximization, and requires to spend more resources on systemic innovation. It is time to think about this.

Acknowledgments   We acknowledge support from SNF (grant The Anatomy of systemic financial risk, No. 162776) and ERC Advanced Investigator Grant MOMENTUM (Grant No. 324247).

References

Asilomar AI Principles. https://futureoflife.org/ai-principles/

Bonnefon, J.-F., Shariff, A., Rahwan, I.: The social dilemma of autonomous vehicles. Science **352**, 1573–1576 (2016)

Bullis, K.: How vehicle automation will cut fuel consumption (24 October 2011). https://www.technologyreview.com/s/425850/how-vehicle-automation-will-cut-fuel-consumption/

Deng, B.: Machine ethics: The robot's dilemma. Nature **523**, 24–26 (2015)

Future of Life Institute: Autonomous weapons: An open letter from AI and robotics researchers (28 July 2015). https://futureoflife.org/open-letter-autonomous-weapons/

Gao, P., Hensley, R., Zielke, A.: A roadmap to the future for the auto industry. McKinsey Quarterly. (October 2014). www.mckinsey.com/industries/automotive-and-assembly/our-insights/a-road-map-to-the-future-for-the-auto-industry

Hackers remotely kill jeep highway. Wired, July 24, 2015. Available at https://www.wired.com/2015/07/hackers-remotely-kill-jeep-highway/

Hackers reveal nasty new car attacks. Forbes, July 24, 2013. Available at https://www.forbes.com/sites/andygreenberg/2013/07/24/hackers-reveal-nasty-new-car-attacks-with-me-behind-the-wheel-video/#45c73d7b228c

Hamburg, F.: Een computermodel voor het ondersteunen van euthanasiebeslissingen (Maklu, 2005)

Helbing, D.: Globally networked risks and how to respond. Nature **497**, 51–59 (2013)

Helbing, D., Pournaras, E.: Build digital democracy. Nature **527**, 33–34 (2015)

IEEE: Ethically Aligned Design Initiative. http://standards.ieee.org/develop/indconn/ec/autono mous_systems.html

Leben, D.: A Rawlsian algorithm for autonomous vehicles. Ethics Inf. Technol. **19**, 107–115 (2017)

Posner, E.A., Sunstein, C.R.: Dollars and death. Univ. Chicago Law Rev. **72**, 537–598 (2005)

Rasmussen, K.B.: Should the probabilities count? Philos. Stud. **159**, 205–218 (2012)

Russell, S., Hauert, S., Altmann, R., Veloso, M.: Robotics: Ethics of artificial intelligence. Nature. **521**, 415–418 (2015)

D. Storm, ACLU: Orwellian Citizen Score, China's credit score system, is a warning for Americans. Computerworld (7 October 2015). https://www.computerworld.com/article/2990203/security/aclu-orwellian-citizen-score-chinas-credit-score-system-is-a-warning-for-americans.html

The Telegraph: Nov 18, 2015. It's official – drivers of luxury cars cause more accidents, insurers say. http://www.telegraph.co.uk/finance/personalfinance/insurance/motorinsurance/11993627/Its-official-drivers-of-luxury-cars-cause-more-accidents-insurers-say.html

van Arem, B., van Driel, C.J.G., Visser, R.: The impact of cooperative adaptive cruise control on traffic-flow characteristics. IEEE Trans. Intell. Transp. Syst. **7**, 429–436 (2006)

Van den Hoven, J., Lokhorst, G.-J., Van de Poel, I.: Engineering and the problem of moral oveerload. Sci. Eng. Ethics **18**(1), 143–155 (2012)

Chapter 6
Societal, Economic, Ethical and Legal Challenges of the Digital Revolution: From Big Data to Deep Learning, Artificial Intelligence, and Manipulative Technologies

Dirk Helbing

In the wake of the on-going digital revolution, we will see a dramatic transformation of our economy and most of our societal institutions. While the benefits of this transformation can be massive, there are also tremendous risks to our society. After the automation of many production processes and the creation of self-driving vehicles, the automation of society is next. This is moving us to a tipping point and to a crossroads: we must decide between a society in which the actions are determined in a top-down way and then implemented by coercion or manipulative technologies (such as personalized ads and nudging) or a society, in which decisions are taken in a free and participatory way and mutually coordinated. Modern information and communication systems (ICT) enable both, but the latter has economic and strategic benefits. The fundaments of human dignity, autonomous decision-making, and democracies are shaking, but I believe that they need to be vigorously defended, as they are not only core principles of livable societies, but also the basis of greater efficiency and success.

This chapter by Dirk Helbing reprints the article Societal, Economic, Ethical and Legal Challenges of the Digital Revolution, published first on 21 May 2015 in Jusletter IT (weblaw.ch) (reprinted with permission).

This document includes and reproduces some paragraphs of the following documents: "Big Data—Zauberstab und Rohstoff des 21. Jahrhunderts" published in Die Volkswirtschaft—Das Magazin für Wirtschaftspolitik (5/2014), see http://www.dievolkswirtschaft.ch/files/editions/201405/pdf/04_Helbing_DE.pdf; for an English translation see chapter 7 of D. Helbing (2015) Thinking Ahead—Essays on Big Data, Digital Revolution, and Participatory Market Society (Springer, Berlin).

D. Helbing (✉)
ETH Zurich, Zürich, Switzerland

TU Delft, Delft, Netherlands

Complexity Science Hub, Vienna, Austria
e-mail: dhelbing@ethz.ch

Those who surrender freedom for security (I would add "efficiency" or "performance" here as well) will not have, nor do they deserve, either one...

Benjamin Franklin

6.1 Overview of Some New Digital Technology Trends

6.1.1 Big Data

In a globalized world, companies and countries are exposed to a harsh competition. This produces a considerable pressure to create more efficient systems—a tendency which is re-inforced by high debt levels.

Big Data seems to be a suitable answer to this. Mining Big Data offers the potential to create new ways to optimize processes, identify interdependencies and make informed decisions. There's no doubt that Big Data creates new business opportunities, not just because of its application in marketing, but also because information itself is becoming monetized.

Technology gurus preach that Big Data is becoming the oil of the twenty-first century, a new commodity that can be tapped for profit. As the virtual currency BitCoin temporarily became more valuable than gold, it can be even literally said that data can be mined into money in a way which would previously have been considered a fairy tale. Although many Big Data sets are proprietary, the consultancy company McKinsey recently estimated that the additional value of Open Data alone amounts to be \$3–5 trillion per year.[1] If the worth of this publicly available information were evenly divided among the world's population, every person on Earth would receive an additional \$700 per year. We now see Open Government initiatives all over the world, aiming to improve services to citizens while having to cut costs. Even the G8 is pushing for Open Data as this is crucial to mobilize the full societal and economic capacity.[2]

The potential of Big Data spans every area of social activity, from the processing of human language and the management of financial assets, to the harnessing of information enabling large cities to manage the balance between energy consumption and production. Furthermore, Big Data holds the promise to help protect our environment, to detect and reduce risks, and to discover opportunities that would otherwise have been missed. In the area of medicine, Big Data could make it possible to tailor medications to patients, thereby increasing their effectiveness and reducing

---

[1]McKinsey and Co. Open data: Unlocking innovation and performance with liquid information, http://www.mckinsey.com/insights/business_technology/open_data_unlocking_innovation_and_performance_with_liquid_information

[2]http://opensource.com/government/13/7/open-data-charter-g8http://ec.europa.eu/digital-agenda/en/news/eu-implementation-g8-open-data-charterhttp://ec.europa.eu/information_society/newsroom/cf/dae/document.cfm?doc_id=3489

their side effects. Big Data could also accelerate the research and development of new drugs and focus resources on the areas of greatest need.

Big Data applications are spreading like wildfire. They facilitate personalized offers, services and products. One of the greatest successes of Big Data is automatic speech recognition and processing. Apple's Siri understands you when asking for a restaurant, and Google Maps can lead you there. Google translate interprets foreign languages by comparing them with a huge collection of translated texts. IBM's Watson computer even understands human language. It can not only beat experienced quiz show players, but take care of customer hotlines and patients—perhaps better than humans. IBM has just decided to invest $1 billion to further develop and commercialize the system.

Of course, Big Data play an important role in the financial sector. Approximately 70% of all financial market transactions are now made by automated trading algorithms. In just one day, the entire money supply of the world is traded. So much money also attracts organized crime. Therefore, financial transactions are scanned by Big Data algorithms for abnormalities to detect suspicious activities. The company Blackrock uses a similar software, called "Aladdin", to successfully speculate with funds amounting approximately to the gross domestic product (GDP) of Europe.

The Big Data approach is markedly different from classical data mining approaches, where datasets have been carefully collected and carefully curated in databases by scientists or other experts. However, each year we now produce as much data as in the entire history of humankind, i.e. in all the years before. This exceeds by far human capacities to curate all data. In just one minute, 700,000 google queries and 500,000 facebook comments are sent. Besides this, enormous amounts of data are produced by all the traces that human activities are now leaving in the Internet. This includes shopping and financial data, geo-positioning and mobility data, social contacts, opinions posted in social networks, files stored in dropbox or some other cloud storage, emails posted or received through free accounts, ebooks read, including time spent on each page and sentences marked, Google or Apple Siri queries asked, youtube or TV movies watched on demand, and games played. Modern game engines and smart home equipment would also sense your activities at home, digital glasses would transmit what you see, and gene data are also massively gathered now.

Meanwhile, the data sets collected by companies such as ebay, Walmart or Facebook, reach the size of petabytes (1 million billion bytes)—one hundred times the information content of the largest library in the world: the U.S. Library of Congress. The mining of Big Data opens up entirely new possibilities for process optimization, the identification of interdependencies, and decision support. However, Big Data also comes with new challenges, which are often characterized by four criteria:

- *volume:* the file sizes and number of records are huge,
- *velocity:* the data evaluation has often to be done in real-time,
- *variety:* the data are often very heterogeneous and unstructured,

- *veracity:* the data are probably incomplete, not representative, and contain errors.

Therefore, completely new algorithms had to be developed, i.e. new computational methods.

6.1.2 Machine Learning, Deep Learning, and Super-Intelligence

To create value from data, it is crucial to turn raw data into useful information and actionable knowledge, some even aim at producing "wisdom" and "clairvoyance" (predictive capabilities). This process requires powerful computer algorithms. Machine learning algorithms do not only watch out for particular patterns, but find patterns even by themselves. This has led Chris Anderson to famously postulate "the end of theory", i.e. the hypothesis that the data deluge makes the scientific method obsolete.[3] If there would be just a big enough quantity of data, machine learning could turn it into high-quality data and come to the right conclusions. This hypothesis has become the credo of Big Data analytics, even though this almost religious belief lacks a proper foundation. I am therefore calling here for a proof of concept, by formulating the following test: Can universal machine learning algorithms, when mining huge masses of experimental physics data, discover the laws of nature themselves, without the support of human knowledge and intelligence?

In spite of these issues, deep learning algorithms are celebrating great successes in everyday applications that do not require an understanding of a hidden logic or causal interdependencies.[4] These algorithms are universal learning procedures which, theoretically, could learn any pattern or input-output relation, given enough time and data. Such algorithms are particularly strong in pattern recognition tasks, i.e. reading, listening, watching, and classifying contents.[5] As a consequence, experts believe that about 50% of all current jobs in the industrial and service sectors will be lost in the next 10–20 years. Moreover, abilities comparable to the human brain are expected to be reached within the next 5–25 years.[6] This has led to a revival of Artificial Intelligence, now often coming under the label "Cognitive Computing".

To be competitive with intelligent machines, humans will in future increasingly need "cognitive assistants". These are digital tools such as Google Now. However, as cognitive assistants get more powerful at exponentially accelerating pace, they

---

[3]Chris Anderson, The End of Theory: The Data Deluge Makes the Scientific Method Obsolete. WIRED Magazine 16.07, http://archive.wired.com/science/discoveries/magazine/16-07/pb_theory

[4]One of the leading experts in this field is Jürgen Schmidhuber.

[5]Jeremy Howard, The wonderful and terrifying implications of computers that can learn, TEDx Brussels, http://www.ted.com/talks/jeremy_howard_the_wonderful_and_terrifying_implications_of_computers_that_can_learn

[6]The point in time when this happens is sometimes called "singularity", according to Ray Kurzweil.

would soon become something like virtual colleagues, then something like digital coaches, and finally our bosses. Robots acting as bosses are already being tested.[7]

Scientists are also working on "biological upgrades" for humans. The first cyborgs, i.e. humans that have been technologically upgraded, already exist. The most well-known of them is Neil Harbisson. At the same time, there is large progress in producing robots that look and behave increasingly like humans. It must be assumed that many science fiction phantasies shown in cinemas and on TV may soon become reality.[8]

Recently, however, there are increasing concerns about artificial super-intelligences, i.e. machines that would be more intelligent than humans. In fact, computers are now better at calculating, at playing chess and most other strategic games, at driving cars, and they are performing many other specialized tasks increasingly well. Certainly, intelligent multi-purpose machines will soon exist.

Only 2 or 3 years ago, most people would have considered it impossible that algorithms, computers, or robots would ever challenge humans as crown of creation. This has changed.[9] Intelligent machines are learning themselves, and it's now conceivable that robots build other robots that are smarter. The resulting evolutionary progress is quickly accelerating, and it is therefore just a matter of time until there are machines smarter than us. Perhaps such super-intelligences already exist. In the following, I am presenting some related quotes of some notable scientists and technology experts, who raise concerns and try to alert the public of the problems we are running into:

For example, Elon Musk of Tesla Motors voiced[10]: "I think we should be very careful about artificial intelligence. If I had to guess at what our biggest existential threat is, it's probably that. So we need to be very careful. ... I am increasingly inclined to think that there should be some regulatory oversight, maybe at the national and international level, just to make sure that we don't do something very foolish..."

Similar critique comes from Nick Bostrom at Oxford University.[11]

Stephen Hawking, the most famous physicist to date, recently said[12]: "Humans who are limited by slow biological evolution couldn't compete and would be superseded. ... The development of full artificial intelligence could spell the end

---

[7]Süddeutsche (11.3.2015) Roboter als Chef, http://www.sueddeutsche.de/leben/roboter-am-arbeitsplatz-billig-freundlich-klagt-nicht-1.2373715

[8]Such movies often serve to familiarize the public with new technologies and realities, and to give them a positive touch (including "Big Brother").

[9]James Barrat (2013) Our Final Invention— Artificial Intelligence and the End of the Human Era (Thomas Dunne Books). Edge Question 2015: What do you think about machines that think? http://edge.org/annual-question/what-do-you-think-about-machines-that-think

[10]http://www.theguardian.com/technology/2014/oct/27/elon-musk-artificial-intelligence-ai-biggest-existential-threat

[11]Nick Bostrom (2014) Superintelligence: Paths, Dangers, Strategies (Oxford University Press).

[12]http://www.bbc.com/news/technology-30290540

of the human race. ... It would take off on its own, and re-design itself at an ever increasing rate."

Furthermore, Bill Gates of Microsoft was quoted[13]: "I am in the camp that is concerned about super intelligence. ... I agree with Elon Musk and some others on this and don't understand why some people are not concerned."

Steve Wozniak, co-founder of Apple, formulated his worries as follows[14]: "Computers are going to take over from humans, no question ... Like people including Stephen Hawking and Elon Musk have predicted, I agree that the future is scary and very bad for people ... If we build these devices to take care of everything for us, eventually they'll think faster than us and they'll get rid of the slow humans to run companies more efficiently ... Will we be the gods? Will we be the family pets? Or will we be ants that get stepped on? I don't know ..."

Personally, I think more positively about artificial intelligence, but I believe that we should engage in distributed collective intelligence rather than creating a few extremely powerful super-intelligences we may not be able to control.[15] It seems that various big IT companies in the Silicon Valley are already engaged in building super-intelligent machines. It was also recently reported that Baidu, the Chinese search engine, wanted to build a "China Brain Project", and was looking for significant financial contributions by the military.[16] Therefore, to be competitive, do we need to sacrifice our privacy for a society-spanning Big Data and Deep Learning project to predict the future of the world? As will become clear later on, I don't think so, because Big Data approaches and the learning of facts from the past are usually bad a predicting fundamental shifts as they occur at societal tipping points, while this is what we mainly need to care about. The combination of explanatory models with little (but right kind of) data is often superior.[17] This can deliver a better description of macro-level societal and economic change, as I will show below, and it's macro-level effects that really matter. Additionally, one should invest in tools that allow one to reveal mechanisms for the management and design of better systems. Such innovative solutions, too, cannot be found by mining data of the past and learning patterns in them.

---

[13]http://www.cnet.com/news/bill-gates-is-worried-about-artificial-intelligence-too/

[14]http://www.washingtonpost.com/blogs/the-switch/wp/2015/03/24/apple-co-founder-on-artificial-intelligence-the-future-is-scary-and-very-bad-for-people/

[15]D. Helbing (2015) Distributed Collective Intelligence: The Network Of Ideas, http://edge.org/response-detail/26194

[16]http://www.scmp.com/lifestyle/technology/article/1728422/head-chinas-google-wants-country-take-lead-developing,   http://www.wantchinatimes.com/news-subclass-cnt.aspx?id=20150307000015&cid=1101

[17]For example, the following approach seems superior to what Google Flu Trends can offer: D. Brockmann and D. Helbing, The hidden geometry of complex, network-driven contagion phenomena. Science 342, 1337–1342 (2013).

6.1.3   *Persuasive Technologies and Nudging to Manipulate Individual Decisions*

Personal data of all kinds are now being collected by many companies, most of which are not well-known to the public. While we surf the Internet, every single click is recorded by cookies, super-cookies and other processes, mostly without our consent. These data are widely traded, even though this often violates applicable laws. By now, there are about 3000–5000 personal records of more or less every individual in the industrialized world. These data make it possible to map the way each person thinks and feels. Their clicks would not only produce a unique finger-print identifying them (perhaps even when surfing anonymously). They would also reveal the political party they are likely to vote for (even though the anonymous vote is an important basis of democracies). Their google searches would furthermore reveal the likely actions they are going to take next (including likely financial trades[18]). There are even companies such as Recorded Future and Palantir that try to predict future individual behavior based on the data available about each of us. Such predictions seem to work pretty well, in more than 90% of all cases. It is often believed that this would eventually make the future course of our society predictable and controllable.

In the past, the attitude was "nobody is perfect, people make mistakes". Now, with the power of modern information technologies, some keen strategists hope that our society could be turned into a perfect clockwork. The feasibility of this approach is already being tested. Personalized advertisement is in fact trying to manipulate people's choices, based on the detailed knowledge of a person, including how he/she thinks, feels, and responds to certain kinds of situations. These approaches become increasingly effective, making use of biases in human decision-making and also subliminal messages. Such techniques address people's subconsciousness, such that they would not necessarily be aware of the reasons causing their actions, similar to acting under hypnosis.

Manipulating people's choices is also increasingly being discussed as policy tool, called "nudging" or "soft paternalism".[19] Here, people's decisions and actions would be manipulated by the state through digital devices to reach certain outcomes, e.g. environmentally friendly or healthier behavior, or also certain election results. Related experiments are being carried out already.[20]

---

[18]T. Preis, H.S. Moat, and H.E. Stanley, Quantifying trading behavior in financial markets using Google Trends. Scientific Reports 3: 1684 (2013).

[19]R.H. Thaler and C.R. Sunstein (2009) Nudge (Penguin Books).

[20]Süddeutsche (11.3.2015) Politik per Psychotrick, http://www.sueddeutsche.de/wirtschaft/verhaltensforschung-am-buerger-politik-per-psychotrick-1.2386755

6.2   Attempt of a Technology Assessment

In the following, I will discuss some of the social, economic, legal, ethical and other implications of the above digital technologies and their use. Like all other technologies, the use of Big Data, Artificial Intelligence, and Nudging can produce potentially harmful side effects, but in this case the impact on our economy and society may be massive. To benefit from the opportunities of digital technologies and minimize their risks, it will be necessary to combine certain technological solutions with social norms and legal regulations. In the following, I attempt to give a number of initial hints, but the discussion below can certainly not give a full account of all issues that need to be addressed.

6.2.1   Problems with Big Data Analytics

The risks of Big Data are manifold. The *security* of digital communication has been undermined. Cybercrime, including data, identity and financial theft, is exploding, now producing an annual damage of the order of three trillion dollars, which is exponentially growing. Critical infrastructures such as energy, financial and communication systems are threatened by cyber attacks. They could, in principle, be made dysfunctional for an extended period of time, thereby seriously disrupting our economy and society. Concerns about cyber wars and digital weapons (D weapons) are quickly growing, as they may be even more dangerous than atomic, biological and chemical (ABC) weapons.

Besides cyber risks, there is a pretty long list of other problems. Results of Big Data analytics are often taken for granted and objective. This is dangerous, because the effectiveness of Big Data is sometimes based more on beliefs than on facts.[21] It is also far from clear that surveillance cameras[22] and predictive policing[23] can really significantly reduce organized and violent crime or that mass surveillance is more

---

[21]For example, many Big Data companies (even big ones) don't make large profits and some are even making losses. Making big money often requires to bring a Big Data company to the stock market, or to be bought by another company.

[22]M. Gill and A. Spriggs: Assessing the impact of CCTV. Home Office Research, Development and Statistics Directorate (2005), https://www.cctvusergroup.com/downloads/file/Martin%20gill.pdf; see also BBC News (August 24, 2009) 1000 cameras 'solve one crime', http://news.bbc.co.uk/2/hi/uk_news/england/london/8219022.stm

[23]Journalist's Resource (November 6, 2014) The effectiveness of predictive policing: Lessons from a randomized controlled trial, http://journalistsresource.org/studies/government/criminal-justice/predictive-policing-randomized-controlled-trial. ZEIT Online (29.3.2015) Predictive Policing—Noch hat niemand bewiesen, dass Data Mining der Polizei hilft, http://www.zeit.de/digital/datenschutz/2015-03/predictive-policing-software-polizei-precobs

effective in countering terrorism than classical investigation methods.[24] Moreover, one of the key examples of the power of Big Data analytics, Google Flu Trends, has recently been found to make poor predictions. This is partly because advertisements bias user behaviors and search algorithms are being changed, such that the results are not stable and reproducible.[25] In fact, Big Data curation and calibration efforts are often low. As a consequence, the underlying datasets are typically not representative and they may contain many errors. Last but not least, Big Data algorithms are frequently used to reveal optimization potentials, but their results may be unreliable or may not reflect any causal relationships. Therefore, conclusions from Big Data are not necessarily correct.

A naive application of Big Data algorithms can easily lead to mistakes and wrong conclusions. The error rate in classification problems (e.g. the distinction between "good" and "bad" risks) is often significant. Issues such as wrong decisions or discrimination are serious problems.[26] In fact, anti-discrimination laws may be implicitly undermined, as results of Big Data algorithms may imply disadvantages for women, handicapped people, or ethnic, religious, and other minorities. This is, because insurance offers, product prices of Internet shops, and bank loans increasingly depend on behavioral variables, and on specifics of the social environment, too. It might happen, for example, that the conditions of a personal loan depend on the behavior of people one has never met. In the past, some banks have even terminated loans, when neighbors have failed to make their payments on time.[27] In other words, as we lose control over our personal data, we are losing control over our lives, too. How will we then be able to take responsibility for our life in the future, if we can't control it any longer?

This brings us to the point of privacy. There are a number of important points to be considered. First of all, surveillance scares people, particularly minorities. All minorities are vulnerable, but the success of our society depends on them (e.g. politicians, entrepreneurs, intellectuals). As the "Volkszählungsurteil"[28] correctly concludes, the continuous and uncontrolled recording of data about individual behaviors is undermining chances of personal, but also societal development.

---

[24]The Washington Post (January 12, 2014) NSA phone record collection does little to prevent terrorist attacks, group says, http://www.washingtonpost.com/world/national-security/nsa-phone-record-collection-does-little-to-prevent-terrorist-attacks-group-says/2014/01/12/8aa860aa-77dd-11e3-8963-b4b654bcc9b2_story.html?hpid=z4; see also http://securitydata.newamerica.net/nsa/analysis

[25]D.M. Lazer et al. The Parable of Google Flu: Traps in Big Data Analytics, Science 343, 1203–1205 (2014).

[26]D. Helbing (2015) Thinking Ahead, Chapter 10 (Springer, Berlin). See also https://www.ftc.gov/news-events/events-calendar/2014/09/big-data-tool-inclusion-or-exclusionhttps://www.whitehouse.gov/issues/technology/big-data-reviewhttps://www.whitehouse.gov/sites/default/files/docs/Big_Data_Report_Nonembargo_v2.pdfhttp://www.wsj.com/articles/SB10001424052702304178104579535970497908560

[27]This problem is related with the method of "geoscoring", see http://www.kreditforum.net/kreditwuerdigkeit-und-geoscoring.html/

[28]http://de.wikipedia.org/wiki/Volksz%C3%A4hlungsurteil

Society needs innovation to adjust to change (such as demographic, environmental, technological or climate change). However, innovation needs a cultural setting that allows to experiment and make mistakes.[29] In fact, many fundamental inventions have been made by accident or even mistake (Porcelain, for example, resulted from attempts to produce gold). A global map of innovation clearly shows that fundamental innovation mainly happens in free and democratic societies.[30] Experimenting is also needed to become an adult who is able to judge situations and take responsible decisions.

Therefore, society needs to be run in a way that is tolerant to mistakes. But today one may get a speed ticket for having been 1 km/h too fast—seriously (at least in Switzerland)! In future, in our over-regulated world, one might get tickets for almost anything.[31] Big Data would make it possible to discover and sanction any small mistake. In the USA, there are already 10 times more people in prison than in Europe (and more than in China and Russia, too). Is this our future, and does it have anything to do with the free society we used to live in? However, if we would punish only a sample of people making mistakes, how would this be compatible with fairness? Wouldn't this end in arbitrariness and undermine justice? And wouldn't the principle of assumed innocence be gone, which is based on the idea that the majority of us are good citizens, and only a few are malicious and to be found guilty? Undermining privacy can't work well. It questions trust in the citizens, and this undermines the citizens' trust in the government, which is the basis of its legitimacy and power. The saying that "trust is good, but control is better" is not entirely correct: control cannot fully replace trust.[32] A well-functioning and efficient society needs a suitable combination of both. "Public" without "private" wouldn't work well. Privacy provides opportunities to explore new ideas and solutions. It helps to recover from the stress of daily adaptation and reduces conflict in a dense population of people with diverse preferences and cultural backgrounds. Public and private are two sides of the same medal. If everything is public, this will eventually undermine social norms.[33] On the long run, the consequence could be a shameless society, or if any deviation from established norms is sanctioned, a totalitarian society.

Therefore, while the effects of mass surveillance and privacy intrusion are not immediately visible, they might still cause a long-term damage by undermining the fabric of our society: social norms and culture. It is highly questionable whether the

---

[29]The Silicon Valley is well-known for this kind of culture.

[30]A. Mazloumian et al. Global multi-level analysis of the 'scientific food web', Scientific Reports 3: 1167 (2013), http://www.nature.com/srep/2013/130130/srep01167/full/srep01167.html? message-global=remove

[31]J. Schmieder (2013) Mit einem Bein im Knast—Mein Versuch, ein Jahr lang gesetzestreu zu leben (Bertelsmann).

[32]Detlef Fetchenhauer, Six reasons why you should be moretrustful, TEDx Groningen, https://www.youtube.com/watch?v=gZlzCc57qX4

[33]A. Diekmann, W. Przepiorka, and H. Rauhut, Lifting the veil of ignorance: An experiment on the contagiousness of norm violations, preprint http://cess.nuff.ox.ac.uk/documents/DP2011/CESS_DP2011_004.pdf

economic benefits would really outweight this, and whether a control-based digital society would work at all. I rather expect such societal experiments to end in disaster.

6.2.2 Problems with Artificial Intelligence and Super-Intelligence[34]

The globalization and networking of our world has caused a level of interdependency and complexity that no individual can fully grasp. This leads to the awkward situation that every one of us sees only part of the picture, which has promoted the idea that we should have artificial super-intelligences that may be able to overlook the entire knowledge of the world. However, learning such knowledge (not just the facts, but also the implications) might progress more slowly than our world changes and human knowledge progresses.[35]

It is also important to consider that the meaning of data depends on context. This becomes particularly clear for ambiguous content. Therefore, like our own brain, an artificial intelligence based on deep learning will sometimes see spurious correlations, and it will probably have some prejudices, too.

Unfortunately, having more information than humans (as cognitive computers have it today) doesn't mean to be objective or right. The problem of "over-fitting", according to which there is a tendency to fit meaningless, random patterns in the data is just one possible issue. The problems of parameter sensitivity or of "chaotic" or "turbulent" system dynamics will restrict possibilities to predict future events, to assess current situations, or even to identify the correct model parameters describing past events.[36] Despite these constraints, a data-driven approach would always deliver some output, but this might be just an "opinion" of an intelligent machine rather than a fact. This becomes clear if we assume to run two identical super-intelligent machines in different places. As they are not fed with exactly the same information, they would have different learning histories, and would sometimes come to different conclusions. So, super-intelligence is no guarantee to find a

---

[34]Note that super-intelligent machines may be seen as an implementation of the concept of the "wise king". However, as I am saying elsewhere, this is not a suitable approach to govern complex societies (see also the draft chapters of my book on the Digital Society at http://www.ssrn.com and https://futurict.blogspot.com, particularly the chapter on the Complexity Time Bomb: http://papers.ssrn.com/sol3/papers.cfm?abstract_id=2502559). Combinatorial complexity must be answered by combinatorial, i.e. collective intelligence, and this needs personal digital assistants and suitable information platforms for coordination.

[35]Remember that it takes about two decades for a human to be ready for responsible, self-determined behavior. Before, however, he/she may do a lot of stupid things (and this may actually happen later, too).

[36]I. Kondor, S. Pafka, and G. Nagy, Noise sensitivity of portfolio selection under various risk measures, Journal of Banking & Finance 31(5), 1545–1573 (2007).

solution that corresponds to the truth.[37] And what if a super-intelligent machine catches a virus and gets something like a "brain disease"?

The greatest problem is that we might be tempted to apply powerful tools such as super-intelligent machines to shape our society at large. As it became obvious above, super-intelligences would make mistakes, too, but the resulting damage might be much larger and even disastrous. Besides this, super-intelligences might emancipate themselves and become uncontrollable. They might also start to act in their own interest, or lie.

Most importantly, powerful tools will always attract people striving for power, including organized criminals, extremists, and terrorists. This is particularly concerning because there is no 100% reliable protection against serious misuse. At the 2015 WEF meeting, Misha Glenny said: "There are two types of companies in the world: those that know they've been hacked, and those that don't".[38] In fact, even computer systems of many major companies, the US military, the Pentagon and the White House have been hacked in the past, not to talk about the problem of data leaks... Therefore, the growing concerns regarding building and using super-intelligences seem to be largely justified.

6.2.3   Problems with Manipulative ("Persuasive") Technologies[39]

The use of information technology is changing our behavior. This fact invites potential misuse, too.[40] Society-scale experiments with manipulative technologies are likely to have serious side effects. In particular, influencing people's decision-

---

[37]It's quite insightful to have two phones talk to each other, using Apple's Siri assistant, see e.g. this video: https://www.youtube.com/watch?v=WuX509bXV_w

[38]http://www.brainyquote.com/quotes/quotes/m/mishaglenn564076.html, see also http://www. businessinsider.com/fbi-director-china-has-hacked-every-big-us-company-2014-10

[39]In other places (http://futurict.blogspot.com/2014/10/crystal-ball-and-magic-wandthe.html), I have metaphorically compared these technologies with a "magic wand" ("Zauberstab"). The problem with these technologies is: they are powerful, but if we don't use them well, their use can end in disaster. A nice poem illustrating this is The Sourcerer's Apprentice by Johann Wolfgang von Goethe: http://germanstories.vcu.edu/goethe/zauber_dual.html, http://www.rither.de/a/deutsch/goethe/der-zauberlehrling/

[40]For example, it recently became public that Facebook had run a huge experiment trying to manipulate people's mood: http://www.theatlantic.com/technology/archive/2014/09/facebooks-mood-manipulation-experiment-might-be-illegal/380717/ This created a big "shit storm": http://www.wsj.com/articles/furor-erupts-over-facebook-experiment-on-users-1404085840. However, it was also attempted to influence people's voting behavior: http://www.nzz.ch/international/kampf-um-den-glaesernen-waehler-1.18501656 OkCupid even tried to manipulate people's private emotions: http://www.theguardian.com/technology/2014/jul/29/okcupid-experiment-human-beings-dating It is also being said that each of our Web searches now triggers about 200 experiments.

making undermines the principle of the "wisdom of crowds",[41] on which democratic decision-making and also the functioning of financial markets is based. For the "wisdom of crowds" to work, one requires sufficiently well educated people who gather and judge information separately and make their decisions independently. Influencing people's decisions will increase the likelihood of mistakes, which might be costly. Moreover, the information basis may get so biased over time that no one, including government institutions and intelligent machines, might be able to make reliable judgments.

Eli Pariser raised a related issue, which he called the "filter bubble". As we increasingly live in a world of personalized information, we are less and less confronted with information that doesn't fit our beliefs and taste. While this creates a feeling to live in the world we like, we will lose awareness of other people's needs and their points of view. When confronted with them, we may fail to communicate and interact constructively. For example, the US political system seems to increasingly suffer from the inability of republicans and democrats to make compromises that are good for the country. When analyzing their political discourse on certain subjects, it turns out that they don't just have different opinions, but they also use different words, such that there is little chance to develop a shared understanding of a problem.[42] Therefore, some modern information systems haven't made it easier to govern a country—on the contrary.

In perspective, manipulative technologies may be seen as attempts to "program" people. Some crowd sourcing techniques such as the services provided by Mechanical Turk come already pretty close to this. Here, people pick up jobs of all kinds, which may just take a few minutes. For example, you may have a 1000 page manual translated in a day, by breaking it down into sufficiently many micro-translation jobs.[43] In principle, however, one could think of anything, and people might not even be aware of the outcome they are jointly producing.[44]

Importantly, manipulation incapacitates people, and it makes them less capable of solving problems by themselves.[45] On the one hand, this means that they increasingly lose control of their judgments and decision-making. On the other hand, who should be held responsible for mistakes that are based on manipulated decisions? The one who took the wrong decision or the one who made him or her take the wrong decision? Probably the latter, particularly as human brains can decreasingly

---

[41]J. Lorenz et al. How social influence can undermine the wisdom of crowd effect, Proceedings of the National Academy of Science of the USA 108 (22), 9020–9025 (2011); see also J. Surowiecki (2005) The Wisdom of Crowds (Anchor).

[42]See Marc Smith's analyses of political discourse with NodeXL: http://nodexl.codeplex.com/

[43]M. Bloodgood and C. Callison-Burch, Using Mechanical Turk to build machine translation evaluation sets, http://www.cis.upenn.edu/ccb/publications/using-mechanical-turk-to-build-machine-translation-evaluation-sets.pdf

[44]In an extreme case, this might even be a criminal act.

[45]Interestingly, for IBM Watson (the intelligent cognitive computer) to work well, it must be fed with non-biased rather than with self-consistent information, i.e. pre-selecting inputs to get rid of contradictory information reduces Watson's performance.

keep up with the performance of computer systems (think, for example, of high-frequency trading).

Finally, we must be aware of another important issue. Some keen strategists believe that manipulative technologies would be perfect tools to create a society that works like a perfect machine. The idea behind this is as follows: A super-intelligent machine would try to figure out an optimal solution to a certain problem, and it would then try to implement it using punishment or manipulation, or both. In this context, one should evaluate again what purposes recent editions of security laws (such as the BÜPF) might be used for, besides fighting true terrorists. It is certainly concerning if people can be put to jail for contents on their computer hard disks, while at the same time hard disks are known to have back doors, and secret services are allowed to download materials to them. This enables serious misuse, but it also questions whether hard disk contents can be still accepted as evidence at court.

Of course, one must ask, whether it would be really possible to run a society by a combination of surveillance, manipulation and coercion. The answer is: probably yes, but given the complexity of our world, I expect this would not work well and not for long. One might therefore say that, in complex societies, the times where a "wise king" or "benevolent dictator" could succeed are gone.[46] But there is the serious danger that some ambitious people might still try to implement the concept and take drastic measures in desperate attempts to succeed. Minorities, who are often seen to produce "unnecessary complexity", would probably get under pressure.[47] This would reduce social, cultural and economic diversity.

As a consequence, this would eventually lead to a socio-economic "diversity collapse", i.e. many people would end up behaving similarly. While this may appear favorable to some people, one must recognize that diversity is the basis of innovation, economic development,[48] societal resilience, collective intelligence, and individual happiness. Therefore, socio-economic and cultural diversity must be protected in a similar way as we have learned to protect biodiversity.[49]

Altogether, it is more appropriate to compare a social or economic system to an ecosystem than to a machine. It then becomes clear that a reduction of diversity corresponds to the loss of biological species in an ecosystem. In the worst case, the ecosystem could collapse. By analogy, the social or economic system would lose performance and become less functional. This is what typically happens in

---

[46]It seems, for example, that the attempts of the world's superpower to extend its powers have rather weakened it: we are now living in a multi-polar world. Coercion works increasingly less. See the draft chapters of my book on the Digital Society at http://ssrn.com for more information.

[47]even though one never knows before what kinds of ideas and social mechanisms might become important in the future—innovation always starts with minorities.

[48]C.A. Hidalgo et al. The product space conditions the development of nations, Science 317, 482–487 (2007). According to Jürgen Mimkes, economic progress (which goes along with an increase in complexity) also drives a transition from autocratic to democratic governance above a certain gross domestic product per capita. In China, this transition is expected to happen soon.

[49]This is the main reason why one should support pluralism.

totalitarian regimes, and it often ends with wars as a result of attempts to counter the systemic instability caused by a diversity collapse.[50]

In conclusion, to cope with diversity, engaging in interoperability is largely superior to standardization attempts. That is why I am suggesting below to develop personal digital assistants that help to create benefits from diversity.

6.3   Recommendations

I am a strong supporter of using digital technologies to create new business opportunities and to improve societal well-being. Therefore, I think one shouldn't stop the digital revolution. (Such attempts would anyway fail, given that all countries are exposed to harsh international competition.) However, like with every technology, there are also potentially serious side effects, and there is a dual use problem.

If we use digital technologies in the wrong way, it could be disastrous for our economy, ending in mass unemployment and economic depression. Irresponsible uses could also be bad for our society, potentially ending (intentionally or not) in more or less totalitarian regimes with little individual freedoms.[51] There are also serious security issues due to exponentially increasing cyber crime, which is partially related to the homogeneity of our current Internet, the lack of barriers (for the sake of efficiency), and the backdoors in many hard- and software systems.

Big Data produces further threats. It can be used to ruin personal careers and companies, but also to launch cyber wars.[52] As we don't allow anyone to own a nuclear bomb or to drive a car without breaks and other safety equipment, we must regulate and control the use of Big Data, too, including the use of Big Data by governments and secret services. This seems to require a sufficient level of transparency, otherwise it is hard to judge for anyone whether we can trust such uses and what are the dangers.

6.3.1   Recommendations Regarding Big Data

The use of Big Data should meet certain quality standards. This includes the following aspects:

---

[50]See the draft chapters of D. Helbing's book on the Digital Society at http://www.ssrn.com, particular the chapter on the Complexity Time Bomb.

[51]One might distinguish these into two types: dictatorships based on surveillance ("Big Brother") and manipulatorships ("Big Manipulator").

[52]As digital weapons, so-called D-weapons, are certainly not less dangerous than atomic, biological and chemical (ABC) weapons, they would require international regulation and control.

- Security issues must be paid more attention to. Sensitive data must be better protected from illegitimate access and use, including hacking of personal data. For this, more and better data encryption might be necessary.
- Storing large amounts of sensitive data in one place, accessible with a single password appears to be dangerous. Concepts such as distributed data storage and processing are advised.
- It should not be possible for a person owning or working in a Big Data company or secret service to look into personal data in unauthorized ways (think of the LoveINT affair, where secret service staff was spying on their partners or ex-partners[53]).
- Informational self-determination (i.e. the control of who uses what personal data for what purpose) is necessary for individuals to keep control of their lives and be able to take responsibility for their actions.
- It should be easy for users to exercise their right of informational self-determination, which can be done by means of Personal Data Stores, as developed by the MIT[54] and various companies. Microsoft seems to be working on a hardware-based solution.
- It must be possible and reasonably easy to correct wrong personal data.
- As Big Data analytics often results in meaningless patterns and spurious correlations, for the sake of objectivity and in order to come to reliable conclusions, it would be good to view its results as hypotheses and to verify or falsify them with different approaches afterwards.
- It must be ensured that scientific standards are applied to the use of Big Data. For example, one should require the same level of significance that is demanded in statistics and for the approval of medical drugs.
- The reproducibility of results of Big Data analytics must be demanded.
- A sufficient level of transparency and/or independent quality control is needed to ensure that quality standards are met.
- It must be guaranteed that applicable antidiscrimination laws are not implicitly undermined and violated.
- It must be possible to challenge and check the results of Big Data analytics.
- Efficient procedures are needed to compensate individuals and companies for improper data use, particularly for unjustified disadvantages.
- Serious violations of constitutional rights and applicable laws should be confronted with effective sanctions.
- To monitor potential misuse and for the sake of transparency, the processing of sensitive data (such as personal data) should probably be always logged.
- Reaping private benefits at the cost of others or the public must be sanctioned.

---

[53] see http://www.washingtonpost.com/blogs/the-switch/wp/2013/08/24/loveint-when-nsa-officers-use-their-spying-power-on-love-interests/

[54] http://openpds.media.mit.edu/

- As for handling dangerous goods, potentially sensitive data operations should require particular qualifications and a track record of responsible behavior (which might be implemented by means of special kinds of reputation systems).

It must be clear that digital technologies will only thrive if they are used in a responsible way. For companies, the trust of consumers and users is important to gain and maintain a large customer base. For governments, public trust is the basis of legitimacy and power. Losing trust would, therefore, cause irrevocable damage.

The current problem is the use of cheap technology. For example, most software is not well tested and not secure. Therefore, Europe should invest in high-quality services of products. Considering the delay in developing data products and services, Europe must anyway find a strategy that differentiates itself from its competitors (which could include an open data and open innovation strategy, too[55]).

Most, if not all functionality currently produced with digital technologies (including certain predictive "Crystal Ball" functionality[56]) can be also obtained in different ways, particularly in ways that are compatible with constitutional and data protection laws[57] (see also the Summary, Conclusion, and Discussion). This may come at higher costs and slightly reduced efficiency, but it might be cheaper overall than risking considerable damage (think of the loss of three trillion dollars by cybercrime each year and consider that this number is still exponentially increasing). Remember also that we have imposed safety requirements on nuclear, chemical, genetic, and other technologies (such as cars and planes) for good reasons. In particular, I believe that we shouldn't (and wouldn't need to) give up the very important principle of informational self-determination in order to unleash the value of personal data. Informational self-control is of key importance to keep democracy, individual freedom, and responsibility for our lives. To reach catalytic and synergy effects, I strongly advise to engage in culturally fitting uses of Information and Communication Technologies (ICT).

In order to avoid slowing down beneficial data applications too much, one might think of continuously increasing standards. Some new laws and regulations might become applicable within 2 or 3 years time, to give companies a sufficiently long time to adjust their products and operation. Moreover, it would be useful to have some open technology standards such that all companies (also small and medium-sized ones) have a chance to meet new requirements with reasonable effort. Requiring a differentiated kind of interoperability could be of great benefit.

---

[55]http://horizon-magazine.eu/article/open-data-could-turn-europe-s-digital-desert-digital-rainforest-prof-dirk-helbing_en.html,   https://ec.europa.eu/digital-agenda/en/growth-jobs/open-innovation

[56]http://www.defenseone.com/technology/2015/04/can-military-make-prediction-machine/109561/

[57]D. Helbing and S. Balietti, From social data mining to forecasting socio-economic crises, Eur. Phys. J Special Topics 195, 3–68 (2011); see also http://www.zeit.de/digital/datenschutz/2015-03/datenschutzverordnung-zweckbindung-datensparsamkeit;   http://www.google.com/patents/US8909546

6.3.2 Recommendations Regarding Machine Learning and Artificial Intelligence

Modern data application go beyond Big Data analytics towards (semi-)automatic systems, which typically offer possibilities for users to control certain system parameters (but sometimes there is just the possibility to turn the automatic or the system off). Autopilot systems, high-frequency trading, and self-driving cars are well-known examples. Would we in future even see an automation of society, including an automated voting by digital agents mirroring ourselves?[58]

Automated or autonomous systems are often not a 100% controllable, as they may operate at a speed that humans cannot compete with. One must also realize that today's artificial intelligent systems are not fully programmed. They learn, and they may therefore behave in ways that have not been tested before. Even if their components would be programmed line by line and would be thoroughly tested without showing any signs of error, the interaction of the system components may lead to unexpected behaviors. For example, this is often the case when a car with sophisticated electronic systems shows surprising behavior (such as suddenly not operating anymore). In fact, unexpected ("emergent") behavior is a typical feature of many complex dynamical systems.

The benefits of intelligent learning systems can certainly be huge. However, we must understand that they will sometimes make mistakes, too, even when automated systems are superior to human task performance. Therefore, one should make a reasonable effort to ensure that mistakes by an automated system are outweighted by its benefits. Moreover, possible damages should be sufficiently small or rare, i.e. acceptable to society. In particular, such damages should not pose any large-scale threats to critical infrastructures, our economy, or our society. As a consequence, I propose the following:

- A legal framework for automated technologies and intelligent machines is necessary. Autonomy needs to come with responsibility, otherwise one may quickly end in anarchy and chaos.
- Companies should be accountable for delivering automated technologies that satisfy certain minimum standards of controllability and for sufficiently educating their users (if necessary).
- The users of automated technologies should be accountable for appropriate efforts to control and use them properly.
- Contingency plans should be available for the case where an automated system gets out of control. It would be good to have a fallback level or plan B that can maintain the functionality of the system at the minimum required performance level.

---

[58]https://www.youtube.com/watch?v=mO-3yVKuDXs,   https://www.youtube.com/watch?v=KgVBob5HIm8

- Insurances and other legal or public mechanisms should be put in place to appropriately and efficiently compensate those who have suffered damage.
- Super-intelligences must be well monitored and should have in-built destruction mechanisms in case they get out of control nevertheless.
- Relevant conclusions of super-intelligent systems should be independently checked (as these could also make mistakes, lie, or act selfishly). This requires suitable verification methods, for example, based on collective intelligence. Humans should still have possibilities to judge recommendations of super-intelligent machines, and to put their suggestions in a historical, cultural, social, economic and ethical perspective.
- Super-intelligent machines should be accessible not only to governing political parties, but also to the opposition (and their respectively commissioned experts), because the discussion about the choice of the goal function and the implication of this choice is inevitable. This is where politics still enters in times of evidence- or science-based decision-making.
- The application of automation should affect sufficiently small parts of the entire system only, which calls for decentralized, distributed, modular approaches and engineered breaking points to avoid cascade effects. This has important implications for the design and management of automated systems, particularly of globally coupled and interdependent systems.[59]
- In order to stay in control, governments must regulate and supervise the use of super-intelligences with the support of qualified experts and independent scientists.[60]

6.3.3 Recommendations Regarding Manipulative Technologies

Manipulative technologies are probably the most dangerous among the various digital technologies discussed in this paper, because we might not even notice the manipulation attempts.

In the past, we lived in an information-poor world. Then, we had enough time to assess the value of information, but we did not always have enough information to decide well. With more information (Web search, Wikipedia, digital maps, etc.) orientation is increasingly easy. Now, however, we are faced with a data deluge and are confronted with so much information that we can't assess and process it all. We are blinded by too much information, and this makes us vulnerable to manipulation. We increasingly need information filters, and the question is, who should produce these information filters? A company? Or the state?

---

[59]Note that the scientific field of complexity science has a large fundus of knowledge how to reach globally coordinated results based on local interactions.

[60]After all, humans have to register, too.

In both cases, this might have serious implications for our society, because the filters would pursue particular interests (e.g. to maximize clicks on ads or to manipulate people in favor of nationalism). In this way, we might get stuck in a "filter bubble".[61] Even if this filter bubble would feel like a golden cage, it would limit our imagination and capacity of innovation. Moreover, mistakes can and will always happen, even if best efforts to reach an optimum outcome are made.

While some problems can be solved well in a centralized fashion (i.e. in a top-down way), some optimization problems are notoriously hard and better solved in a distributed way. Innovation is one of these areas.[62] The main problem is that *the most fundamental question of optimization is unsolved, namely what goal function to choose.* When a bad goal function is chosen, this will have bad outcomes, but we may notice this only after many years. As mistakes in choosing the goal function will surely sometimes happen, it could end in disaster when everyone applies the same goal function.[63]

Therefore, one should apply something like a portfolio strategy. Under strongly variable and hardly predictable conditions a diverse strategy works best. Therefore, pluralistic information filtering is needed. In other words, customers, users, and citizens should be able to create, select, share and adapt the information filters they use, thereby creating an evolving ecosystem of increasingly better filters. In fact, everyone would probably be using several different filters (for example, "What's currently most popular?", "What's most controversial?", "What's trendy in my peer group?", "Surprise me!").[64] In contrast, if we leave it to a company or the state to decide how we see the world, we might happen to end up with biased views, and this could lead to terrible mistakes. This could, for example, undermine the "wisdom of crowds", which is currently the basis of free markets and democracies (with benefits such as a high level of performance [not necessarily growth], quality of life, and the avoidance of mistakes such as wars among each other).

In a world characterized by information overload, unbiased and reliable information becomes ever more important. Otherwise the number of mistakes will probably increase. For the digital society to succeed, we must therefore take safeguards against information pollution and biases. Reputation systems might be a suitable instrument, if enough information providers compete efficiently with each other for providing more reliable and more useful information. Additionally, legal sanctions might be necessary to counter intentionally misleading information.

---

[61] E. Pariser (2012) The Filter Bubble: How the New Personalized Web Is Changing What We Read and How We Think (Penguin).

[62] Some problems are so hard that no government and no company in the world have solved them (e.g. how to counter climate change). Large multi-national companies are often surprisingly weak in delivering fundamental innovations (probably because they are too controlling). That's why they keep buying small and medium-sized companies to compensate for this problem.

[63] Similar problems are known for software products that are used by billions of people: a single software bug can cause large-scale problems—and the worrying vulnerability to cyber attacks is further increasing.

[64] We have demonstrated such an approach in the Virtual Journal platform (http://vijo.inn.ac).

Consequently, advertisements should be marked as such, and the same applies to manipulation attempts such as nudging. In other words, the user, customer or citizen must be given the possibility to consciously decide for or against a certain decision or action, otherwise individual autonomy and responsibility are undermined. Similarly as customers of medical drugs are warned of potential side effects, one should state something like "This product is manipulating your decisions and is trying to make you behave in a more healthy way (or in a environmentally friendly way, or whatever it tries to achieve. . .)". The customer would then be aware of the likely effects of the information service and could actively decide whether he or she wants this or not.

Note, however, that it is currently not clear what the side effects of incentivizing the use of manipulative technologies would be. If applied on a large scale, it might be almost as bad as hidden manipulation. Dangerous herding effects might occur (including mass psychology as it occurs in hypes, stock market bubbles, unhealthy levels of nationalism, or the particularly extreme form it took during the Third Reich). Therefore,

- manipulation attempts should be easily recognizable, e.g. by requiring everyone to mark the kind of information (advertisement, opinion, or fact),
- it might be useful to monitor manipulation attempts and their effects,
- the effect size of manipulation attempts should be limited to avoid societal disruptions,
- one should have a possibility to opt out for free from the exposure to manipulative influences,
- measures to ensure pluralism and socio-economic diversity should be required,
- sufficiently many independent information providers with different goals and approaches would be needed to ensure an effective competition for more reliable information services,
- for collective intelligence to work, having a knowledge base of trustable and unbiased facts is key, such that measures against information pollution are advised.[65]

Ethical guidelines, demanding certain quality standards, and sufficient transparency might also be necessary. Otherwise, the large-scale application of manipulative technologies could intentionally or unintentionally undermine the individual freedom of decision-making and the basis of democracies, particularly when nudging techniques become highly effective and are used to manipulate public opinion at large.[66]

---

[65]In fact, to avoid mistakes, the more we are flooded with information the more must we be able to rely on it, as we have increasingly less time to judge its quality.

[66]This could end up in a way of organizing our society that one could characterize as "Big Manipulator" (to be distinguished from "Big Brother").

6.4 Summary, Conclusions and Discussion

Digital technologies offer great benefits, but also substantial risks. They may help us to solve some long-standing problems, but they may also create new and even bigger issues. In particular, if wrongly used, individual autonomy and freedom, responsible decision-making, democracy and the basis of our legal system are at stake. The foundations on which our society is build might be damaged intentionally or unintentionally within a very short time period, which may not give us enough opportunities to prepare for or respond to the challenges.

Currently, some or even most Big Data practices violate applicable data protection laws. Of course, laws can be changed, but some uses of Big Data are also highly dangerous, and incompatible with our constitution and culture. These challenges must be addressed by a combination of technological solutions (such as personal data stores), legal regulations, and social norms. Distributed data, distributed systems and distributed control, sufficiently many competitors and suitably designed reputation systems might be most efficient to avoid misuses of digital technologies, but transparency must be increased as well.

Even though our economy and society will change in the wake of the digital revolution, we must find a way that is consistent with our values, culture, and traditions, because this will create the largest synergy effects. In other words, a China or Singapore model is unlikely to work well in Europe.[67] We must take the next step in our cultural, economic and societal evolution.

I am convinced that it is now possible to use digital technologies in ways that bring the perspectives of science, politics, business, society, cultural traditions, ethics, and perhaps even religion together.[68] Specifically, I propose to use the Internet of Things as basis for a participatory information system called the Planetary Nervous System or Nervousnet, to support tailored measurements, awareness, coordination, collective intelligence, and informational self-determination.[69] The system I suggest would have a resilient systems design and could be imagined as a huge catalyst of socio-economic value generation. It would also support real-time feedbacks through a multi-dimensional exchange system ("multi-dimensional finance"). This approach would allow one to massively increase the efficiency of many systems, as it would support the self-organization of structures, properties and

---

[67]The following recent newspaper articles support this conclusion:http://www.zeit.de/politik/ausland/2015-03/china-wachstum-fuenf-vor-acht,http://bazonline.ch/wirtschaft/konjunktur/China-uebernimmt-die-rote-Laterne/story/20869017, http://www.nzz.ch/international/asien-und-pazifik/singapurer-zeitrechnung-ohne-lee-kuan-yew-1.18510938. In fact, based on a statistical analysis of Jürgen Mimkes and own observations, I expect that China will now undergo a major transformation towards a more democratic state in the coming years. First signs of instability of the current autocratic system are visible already, such as the increased attempts to control information flows.

[68]D. Helbing, Responding to complexity in socio-economic systems: How to build a smart and resilient society? Preprint http://papers.ssrn.com/sol3/papers.cfm?abstract_id=2583391

[69]D. Helbing, Creating ("Making") a Planetary Nervous System as Citizen Web, http://futurict.blogspot.jp/2014/09/creating-making-planetary-nervous.html

functions that we would like to have, based on local interactions. The distributed approach I propose is consistent with individual autonomy, free decision-making, the democratic principle of participation, as well as free entrepreneurial activities and markets. In fact, wealth is not only created by producing economies of scale (i.e. cheap mass production), but also by engaging in social interaction (that's why cities are drivers of the economy[70]).

The proposed approach would also consider (and potentially trade) externalities, thereby supporting other-regarding and fair solutions, which would be good for our environment, too. Finally, everyone could reap the benefits of diversity by using personal digital assistants, which would support coordination and cooperation of diverse actors and reducing conflict.

In conclusion, we have the choice between two kinds of a digital society: (1) a society in which people are expected to obey and perform tasks like a robot or a gearwheel of a perfect machine, characterized by top-down control, limitations of freedom and democracy, and potentially large unemployment rates; (2) a participatory society with space for humans with sometimes surprising behaviors characterized by autonomous but responsible decision-making supported by personal digital assistants, where information is opened up to everyone's benefits in order to reap the benefits of diversity, creativity, and exponential innovation. What society would you choose.

The FuturICT community (www.futurict.eu) has recently worked out a framework for a smart digital society, which is oriented at international leadership, economic prosperity, social well-being, and societal resilience, based on the well-established principle of subsidiarity. With its largely distributed, decentralized approach, it is designed to cope with the complexity of our globalized world and benefit from it.[71]

The FuturICT approach takes the following insights into account:

- Having and using more data is not always better (e.g. due to the problem of "overfitting", which makes conclusions less useful).[72]
- Information always depends on context (and missing context), and it is therefore never objective. One person's signal may be another person's noise and vice versa. It all depends on the question and perspective.[73]

---

[70]L.M.A. Bettencourt et al. Growth, innovation, scaling, and the pace of life in cities, Proceedings of the National Academy of Sciences of the USA 104, 7301–7306 (2007).

[71]See D. Helbing, Globally networked risks and how to respond. Nature 497, 51–59 (2013). Due to the problem of the Complexity Time Bomb (http://papers.ssrn.com/sol3/papers.cfm? abstract_id=2502559), we must either decentralize our world, or it will most likely fragment, i.e. break into pieces, sooner or later.

[72]Having a greater haystack does not make it easier to find a needle in it.

[73]This is particularly well-known for the problem of ambiguity. For example, a lot of jokes are based on this principle.

- Even if individual decisions can be correctly predicted in 96% of all cases, this does not mean that the macro-level outcome would be correctly predicted.[74] This surprising discovery applies to cases of unstable system dynamics, where minor variations can lead to completely different outcomes.[75]
- In complex dynamical systems with many interacting components, even the perfect knowledge of all individual component properties does not necessarily allow one to predict what happens if components interact.[76]
- What governments really need to pay attention to are macro-effects, not micro-behavior. However, the macro-dynamics can often be understood by means of models that are based on aggregate variables and parameters.
- What matters most is whether a system is stable or unstable. In case of stability, variations in individual behavior do not make a significant difference, i.e. we don't need to know what the individuals do. In case of instability, random details matter, such that the predictability is low, and even in the unlikely case that one can exactly predict the course of events, one may not be able to control it because of cascade-effects in the system that exceed the control capacities.[77]
- Surprises and mistakes will always happen. This can disrupt systems, but many inventions wouldn't exist, if this wasn't the case.[78]
- Our economy and society should be organized in a way that manages to keep disruptions small and to respond flexibly to surprises of all kinds. Socio-

---

[74]M. Maes and D. Helbing, Noise can improve social macro-predictions when micro-theories fail, preprint.

[75]We know this also from so-called "phantom traffic jams", which appear with no reason, when the car density exceeds a certain critical value beyond which traffic flow becomes unstable. Such phantom traffic jams could not be predicted at all by knowing all drivers thoughts and feelings in detail. However, they can be understood for example with macro-level models that do not require micro-level knowledge. These models also show how traffic congestion can be avoided: by using driver assistance systems that change the interactions between cars, using real-time information about local traffic conditions. Note that this is a distributed control strategy.

[76]Assume one knows the psychology of two persons, but then they accidentally meet and fall in love with each other. This incident will change their entire lives, and in some cases it will change history too (think of Julius Caesar and Cleopatra, for example, but there are many similar cases). A similar problem is known from car electronics: even if all electronic components have been well tested, their interaction often produces unexpected outcomes. In complex systems, such unexpected, "emergent" system properties are quite common.

[77]In case of cascade effects, a local problem will cause other problems before the system recovers from the initial disruption. Those problems trigger further ones, etc. Even hundreds of policemen could not avoid phantom traffic jams from happening, and in the past even large numbers of security forces have often failed to prevent crowd disasters (they have sometimes even triggered or deteriorated them while trying to avoid them), see D. Helbing and P. Mukerji, Crowd disasters as systemic failures: Analysis of the Love Parade disaster, EPJ Data Science 1:7 (2012).

[78]I am personally convinced that the level of randomness and unpredictability in a society is relatively high, because it creates a lot of personal and societal benefits, such as creativity and innovation. Also think of the success principle of serendipity.

economic systems should be able to resist shocks and recover from them quickly and well. This is best ensured by a resilient system design.[79]

- A more intelligent machine is not necessarily more useful. Distributed collective intelligence can better respond to the combinatorial complexity of our world.[80]
- In complex dynamical systems which vary a lot, are hard to predict and cannot be optimized in real-time (as it applies to NP-hard control problems such as traffic light optimization), distributed control can outperform top-down control attempts by flexibly adapting to local conditions and needs.
- While distributed control may be emulated by centralized control, a centralized approach might fail to identify the variables that matter.[81] Depending on the problem, centralized control is also considerably more expensive, and it tends to be less efficient and effective.[82]
- Filtering out information that matters is a great challenge. Explanatory models that are combined with little, but the right kind of data are best to inform decision-makers. Such models also indicate what kind of data is needed.[83] Finding the right models typically requires interdisciplinary collaborations, knowledge about complex systems, and open scientific discussions that take all relevant perspectives on board.
- Diversity and complexity are not our problem. They come along with the socio-economic and cultural evolution. However, we have to learn how to use complexity and diversity to our advantage. This requires the understanding of the hidden forces behind socio-economic change, the use of (guided) self-organization and digital assistants to create interoperability and to support the coordination of actors with diverse interests and goals.

---

[79]D. Helbing et al. FuturICT: Participatory computing to understand and manage our complex world in a more sustainable and resilient way. Eur. Phys. J. Special Topics 214, 11–39 (2012).

[80]As we know, intellectual discourse can be a very effective way of producing new insights and knowledge.

[81]Due to the data deluge, the existing amounts of data increasingly exceed the processing capacities, which creates a "flashlight effect": while we might look at anything, we need to decide what data to look at, and other data will be ignored. As a consequence, we often overlook things that matter. While the world was busy fighting terrorism in the aftermath of September 11, it did not see the financial crisis coming. While it was focused on this, it did not see the Arab Spring coming. The crisis in Ukraine came also as a surprise, and the response to Ebola came half a year late. Of course, the possibility or likelihood of all these events was reflected by some existing data, but we failed to pay attention to them.

[82]The classical telematics solutions based on a control center approach haven't improved traffic much. Today's solutions to improve traffic flows are mainly based on distributed control approaches: self-driving cars, intervehicle communication, car-to-infrastructure communication etc.

[83]This approach corresponds exactly how Big Data are used at the elementary particle accelerator CERN; 99.9% of measured data are deleted immediately. One only keeps data that are required to answer a certain question, e.g. to validate or falsify implications of a certain theory.

- To catalyze the best outcomes and create synergy effects, information systems should be used in a culturally fitting way.[84]
- Responsible innovation, trustable systems and a sufficient level of transparency and democratic control can be highly beneficial.

As a consequence of the above insights, to reap the benefits of data, I believe we do not need to end privacy and informational self-determination. The best use of information systems is made, if they boost our society and economy to full capacity, i.e. if they use the knowledge, skills, and resources of everyone in the best possible way. This is of strategic importance and requires suitably designed participatory information systems, which optimally exploit the special properties of information.[85] In fact, the value of participatory systems, as pointed out by Jeremy Rifkin[86] and others,[87] becomes particularly clear if we think of the great success of crowd sourcing (Wikipedia, OpenStreetMap, Github, etc.), crowd funding, citizen science and collective ("swarm") intelligence. So, let's build these systems together. What are we waiting for?

---

[84]J. van den Hoven et al. FuturICT—The road towards ethical ICT, Eur. Phys. J. Special Topics 214, 153–181 (2012).

[85]This probably requires different levels of access depending on qualification, reputation, and merit.

[86]J. Rifkin (2013) The Third Industrial Revolution (Palgrave Macmillan Trade); J. Rifkin (2014) The Zero Marginal Cost Society (Palgrave Macmillan Trade).

[87]Government 3.0 initiative of the South Korean government, http://www.negst.com.ng/docu ments/Governing_through_Networks/3-icegov2013_submission_19.pdfhttp://www.koreaittimes. com/story/32400/government-30-future-opening-sharing-communication-and-collaboration

Chapter 7
Will Democracy Survive Big Data and Artificial Intelligence?

Dirk Helbing, Bruno S. Frey, Gerd Gigerenzer, Ernst Hafen, Michael Hagner, Yvonne Hofstetter, Jeroen van den Hoven, Roberto V. Zicari, and Andrej Zwitter

This article by Dirk Helbing, **Bruno S. Frey, Gerd Gigerenzer, Ernst Hafen, Michael Hagner, Yvonne Hofstetter, Jeroen van den Hoven, Roberto V. Zicari, Andrej Zwitter** was first published in Scientific American on February 25, 2017 under the URL https://www.scientificamerican.com/article/will-democracy-survive-big-data-and-artificial-intelligence/. The original German version appeared first in 2015 in Spektrum der Wissenschaft as "DigitalManifest" under the title "Digitale Demokratie statt Datendiktatur". It is accessible via the URL http://www.spektrum.de/thema/das-digital-manifest/1375924

D. Helbing (✉)
ETH Zurich, Zürich, Switzerland

TU Delft, Delft, Netherlands

Complexity Science Hub, Vienna, Austria
e-mail: dhelbing@ethz.ch

B. S. Frey
University of Basel, Basel, Switzerland

G. Gigerenzer
Max Planck Institute for Human Development, Berlin, Germany

E. Hafen · M. Hagner
ETH Zurich, Zürich, Switzerland

Y. Hofstetter
Teramark Technologies, Zolling, Germany

J. van den Hoven
TU Delft, Delft, Netherlands
e-mail: m.j.vandenhoven@tudelft.nl

R. V. Zicari
Goethe University Frankfurt/Main, Frankfurt, Germany

A. Zwitter
University of Groningen, Groningen, Netherlands

© Springer International Publishing AG, part of Springer Nature 2019   73
D. Helbing (ed.), *Towards Digital Enlightenment*,
https://doi.org/10.1007/978-3-319-90869-4_7

We are in the middle of a technological upheaval that will transform the way society is organized. We must make the right decisions now.

Enlightenment is man's emergence from his self-imposed immaturity. Immaturity is the inability to use one's understanding without guidance from another.
—Immanuel Kant, "What is Enlightenment?" (1784)

The digital revolution is in full swing. How will it change our world? The amount of data we produce doubles every year. In other words: in 2016 we produced as much data as in the entire history of humankind through 2015. Every minute we produce hundreds of thousands of Google searches and Facebook posts. These contain information that reveals how we think and feel. Soon, the things around us, possibly even our clothing, also will be connected with the Internet. It is estimated that in 10 years' time there will be 150 billion networked measuring sensors, 20 times more than people on Earth. Then, the amount of data will double every 12 hours. Many companies are already trying to turn this Big Data into Big Money.

Everything will become intelligent; soon we will not only have smart phones, but also smart homes, smart factories and smart cities. Should we also expect these developments to result in smart nations and a smarter planet?

The field of artificial intelligence is, indeed, making breathtaking advances. In particular, it is contributing to the automation of data analysis. Artificial intelligence is no longer programmed line by line, but is now capable of learning, thereby continuously developing itself. Recently, Google's DeepMind algorithm taught itself how to win 49 Atari games. Algorithms can now recognize handwritten language and patterns almost as well as humans and even complete some tasks better than them. They are able to describe the contents of photos and videos. Today 70% of all financial transactions are performed by algorithms. News content is, in part, automatically generated. This all has radical economic consequences: in the coming 10–20 years around half of today's jobs will be threatened by algorithms. 40% of today's top 500 companies will have vanished in a decade.

It can be expected that supercomputers will soon surpass human capabilities in almost all areas—somewhere between 2020 and 2060. Experts are starting to ring alarm bells. Technology visionaries, such as Elon Musk from Tesla Motors, Bill Gates from Microsoft and Apple co-founder Steve Wozniak, are warning that super-intelligence is a serious danger for humanity, possibly even more dangerous than nuclear weapons.

7.1 Is This Alarmism?

One thing is clear: the way in which we organize the economy and society will change fundamentally. We are experiencing the largest transformation since the end of the Second World War; after the automation of production and the creation of self-driving cars the automation of society is next. With this, society is at a crossroads, which promises great opportunities, but also considerable risks. If we take the wrong decisions it could threaten our greatest historical achievements.

In the 1940s, the American mathematician Norbert Wiener (1894–1964) invented cybernetics. According to him, the behavior of systems could be controlled by the means of suitable feedbacks. Very soon, some researchers imagined controlling the economy and society according to this basic principle, but the necessary technology was not available at that time.

Today, Singapore is seen as a perfect example of a data-controlled society. What started as a program to protect its citizens from terrorism has ended up influencing economic and immigration policy, the property market and school curricula. China is taking a similar route. Recently, Baidu, the Chinese equivalent of Google, invited the military to take part in the China Brain Project. It involves running so-called deep learning algorithms over the search engine data collected about its users. Beyond this, a kind of social control is also planned. According to recent reports, every Chinese citizen will receive a so-called "Citizen Score", which will determine under what conditions they may get loans, jobs, or travel visa to other countries. This kind of individual monitoring would include people's Internet surfing and the behavior of their social contacts (see "Spotlight on China").

With consumers facing increasingly frequent credit checks and some online shops experimenting with personalized prices, we are on a similar path in the West. It is also increasingly clear that we are all in the focus of institutional surveillance. This was revealed in 2015 when details of the British secret service's "Karma Police" program became public, showing the comprehensive screening of everyone's Internet use. Is Big Brother now becoming a reality?

7.2   Programmed Society, Programmed Citizens

Everything started quite harmlessly. Search engines and recommendation platforms began to offer us personalised suggestions for products and services. This information is based on personal and meta-data that has been gathered from previous searches, purchases and mobility behaviour, as well as social interactions. While officially, the identity of the user is protected, it can, in practice, be inferred quite easily. Today, algorithms know pretty well what we do, what we think and how we feel—possibly even better than our friends and family or even ourselves. Often the recommendations we are offered fit so well that the resulting decisions feel as if they were our own, even though they are actually not our decisions. In fact, we are being remotely controlled ever more successfully in this manner. The more is known about us, the less likely our choices are to be free and not predetermined by others.

But it won't stop there. Some software platforms are moving towards "persuasive computing." In the future, using sophisticated manipulation technologies, these platforms will be able to steer us through entire courses of action, be it for the execution of complex work processes or to generate free content for Internet platforms, from which corporations earn billions. *The trend goes from programming computers to programming people.*

These technologies are also becoming increasingly popular in the world of politics. Under the label of "nudging," and on massive scale, governments are trying to steer citizens towards healthier or more environmentally friendly behaviour by means of a "nudge"—a modern form of paternalism. The new, caring government is not only interested in what we do, but also wants to make sure that we do the things that it considers to be right. The magic phrase is "big nudging", which is the combination of big data with nudging. To many, this appears to be a sort of digital scepter that allows one to govern the masses efficiently, without having to involve citizens in democratic processes. Could this overcome vested interests and optimize the course of the world? If so, then citizens could be governed by a data-empowered "wise king", who would be able to produce desired economic and social outcomes almost as if with a digital magic wand.

7.3 Pre-programmed Catastrophes

But one look at the relevant scientific literature shows that attempts to control opinions, in the sense of their "optimization", are doomed to fail because of the complexity of the problem. The dynamics of the formation of opinions are full of surprises. Nobody knows how the digital magic wand, that is to say the manipulative nudging technique, should best be used. What would have been the right or wrong measure often is apparent only afterwards. During the German swine flu epidemic in 2009, for example, everybody was encouraged to go for vaccination. However, we now know that a certain percentage of those who received the immunization were affected by an unusual disease, narcolepsy. Fortunately, there were not more people who chose to get vaccinated!

Another example is the recent attempt of health insurance providers to encourage increased exercise by handing out smart fitness bracelets, with the aim of reducing the amount of cardiovascular disease in the population; but in the end, this might result in more hip operations. In a complex system, such as society, an improvement in one area almost inevitably leads to deterioration in another. Thus, large-scale interventions can sometimes prove to be massive mistakes.

Regardless of this, criminals, terrorists and extremists will try and manage to take control of the digital magic wand sooner or later—perhaps even without us noticing. Almost all companies and institutions have already been hacked, even the Pentagon, the White House, and the NSA.

A further problem arises when adequate transparency and democratic control are lacking: the erosion of the system from the inside. Search algorithms and recommendation systems can be influenced. Companies can bid on certain combinations of words to gain more favourable results. Governments are probably able to influence the outcomes too. During elections, they might nudge undecided voters towards supporting them—a manipulation that would be hard to detect. Therefore, whoever controls this technology can win elections—by nudging themselves to power.

This problem is exacerbated by the fact that, in many countries, a single search engine or social media platform has a predominant market share. It could decisively influence the public and interfere with these countries remotely. Even though the European Court of Justice judgment made on 6 October 2015 limits the unrestrained export of European data, the underlying problem still has not been solved within Europe, and even less so elsewhere.

What undesirable side effects can we expect? In order for manipulation to stay unnoticed, it takes a so-called resonance effect—suggestions that are sufficiently customized to each individual. In this way, local trends are gradually reinforced by repetition, leading all the way to the "filter bubble" or "echo chamber effect": in the end, all you might get is your own opinions reflected back at you. This causes social polarization, resulting in the formation of separate groups that no longer understand each other and find themselves increasingly at conflict with one another. In this way, personalized information can unintentionally destroy social cohesion. This can be currently observed in American politics, where Democrats and Republicans are increasingly drifting apart, so that political compromises become almost impossible. The result is a fragmentation, possibly even a disintegration, of society.

Owing to the resonance effect, a large-scale change of opinion in society can be only produced slowly and gradually. The effects occur with a time lag, but, also, they cannot be easily undone. It is possible, for example, that resentment against minorities or migrants get out of control; too much national sentiment can cause discrimination, extremism and conflict.

Perhaps even more significant is the fact that manipulative methods change the way we make our decisions. They override the otherwise relevant cultural and social cues, at least temporarily. In summary, the large-scale use of manipulative methods could cause serious social damage, including the brutalization of behavior in the digital world. Who should be held responsible for this?

7.4 Legal Issues

This raises legal issues that, given the huge fines against tobacco companies, banks, IT and automotive companies over the past few years, should not be ignored. But which laws, if any, might be violated? First of all, it is clear that manipulative technologies restrict the freedom of choice. If the remote control of our behaviour worked perfectly, we would essentially be digital slaves, because we would only execute decisions that were actually made by others before. Of course, manipulative technologies are only partly effective. Nevertheless, our freedom is disappearing slowly, but surely—in fact, slowly enough that there has been little resistance from the population, so far.

The insights of the great enlightener Immanuel Kant seem to be highly relevant here. Among other things, he noted that a state that attempts to determine the happiness of its citizens is a despot. However, the right of individual self-development can only be exercised by those who have control over their lives, which presupposes informational self-determination. This is about nothing less than our most important constitutional rights. A democracy cannot work well unless those rights are respected. If they are constrained, this undermines our constitution, our society and the state.

As manipulative technologies such as big nudging function in a similar way to personalized advertising, other laws are affected too. Advertisements must be marked as such and must not be misleading. They are also not allowed to utilize certain psychological tricks such as subliminal stimuli. This is why it is prohibited to show a soft drink in a film for a split-second, because then the advertising is not consciously perceptible while it may still have a subconscious effect. Furthermore, the current widespread collection and processing of personal data is certainly not compatible with the applicable data protection laws in European countries and elsewhere.

Finally, the legality of personalized pricing is questionable, because it could be a misuse of insider information. Other relevant aspects are possible breaches of the principles of equality and non-discrimination—and of competition laws, as free market access and price transparency are no longer guaranteed. The situation is comparable to businesses that sell their products cheaper in other countries, but try to prevent purchases via these countries. Such cases have resulted in high punitive fines in the past.

Personalized advertising and pricing cannot be compared to classical advertising or discount coupons, as the latter are non-specific and also do not invade our privacy with the goal to take advantage of our psychological weaknesses and knock out our critical thinking.

Furthermore, let us not forget that, in the academic world, even harmless decision experiments are considered to be experiments with human subjects, which would have to be approved by a publicly accountable ethics committee. In each and every case the persons concerned are required to give their informed consent. In contrast, a single click to confirm that we agree with the contents of a hundred-page "terms of use" agreement (which is the case these days for many information platforms) is woefully inadequate.

Nonetheless, experiments with manipulative technologies, such as nudging, are performed with millions of people, without informing them, without transparency and without ethical constraints. Even large social networks like Facebook or online dating platforms such as OkCupid have already publicly admitted to undertaking these kinds of social experiments. If we want to avoid irresponsible research on humans and society (just think of the involvement of psychologists in the torture scandals of the recent past), then we urgently need to impose high standards, especially scientific quality criteria and a code of conduct similar to the Hippocratic Oath. Has our thinking, our freedom, our democracy been hacked?

Let us suppose there was a super-intelligent machine with godlike knowledge and superhuman abilities: would we follow its instructions? This seems possible. But if we did that, then the warnings expressed by Elon Musk, Bill Gates, Steve Wozniak,

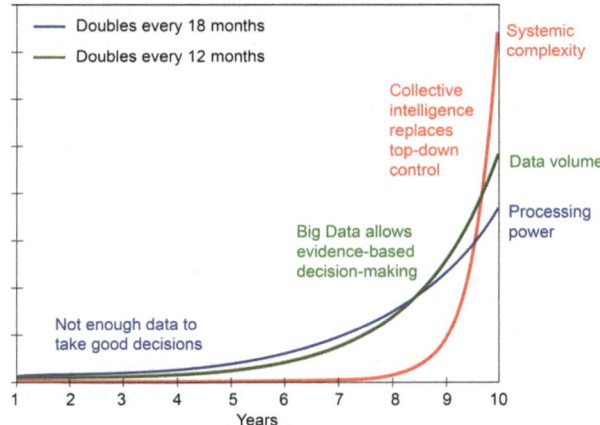

Fig. 7.1 Digital growth. Thanks to Big Data, we can now take better, evidence-based decisions. However, the principle of top-down control increasingly fails, since the complexity of society grows in an explosive way as we go on networking our world. Distributed control approaches will become ever more important. Only by means of collective intelligence will it be possible to find appropriate solutions to the complexity challenges of our world

Stephen Hawking and others would have become true: computers would have taken control of the world. We must be clear that a super-intelligence could also make mistakes, lie, pursue selfish interests or be manipulated. Above all, it could not be compared with the distributed, collective intelligence of the entire population.

The idea of replacing the thinking of all citizens by a computer cluster would be absurd, because that would dramatically lower the diversity and quality of the solutions achievable. It is already clear that the problems of the world have not decreased despite the recent flood of data and the use of personalized information—on the contrary! World peace is fragile. The long-term change in the climate could lead to the greatest loss of species since the extinction of dinosaurs. We are also far from having overcome the financial crisis and its impact on the economy. Cyber-crime is estimated to cause an annual loss of three trillion dollars. States and terrorists are preparing for cyberwarfare.

In a rapidly changing world a super-intelligence can never make perfect decisions (see Fig. 7.1: systemic complexity is increasing faster than data volumes, which are growing faster than the ability to process them, and data transfer rates are limited. This results in disregarding local knowledge and facts, which are important to reach good solutions. Distributed, local control methods are often superior to centralized approaches, especially in complex systems whose behaviors are highly variable, hardly predictable and not capable of real-time optimization. This is already true for traffic control in cities, but even more so for the social and economic systems of our highly networked, globalized world.

Furthermore, there is a danger that the manipulation of decisions by powerful algorithms undermines the basis of "collective intelligence," which can flexibly adapt to the challenges of our complex world. For collective intelligence to work, information searches and decision-making by individuals must occur independently. If our judgments and decisions are predetermined by algorithms, however, this truly leads to a brainwashing of the people. Intelligent beings are downgraded to mere receivers of commands, who automatically respond to stimuli.

In other words: personalized information builds a "filter bubble" around us, a kind of digital prison for our thinking. How could creativity and thinking "out of the box" be possible under such conditions? Ultimately, a centralized system of technocratic behavioral and social control using a super-intelligent information system would result in a new form of dictatorship. Therefore, the top-down controlled society, which comes under the banner of "liberal paternalism," is in principle nothing else than a totalitarian regime with a rosy cover.

In fact, big nudging aims to bring the actions of many people into line, and to manipulate their perspectives and decisions. This puts it in the arena of propaganda and the targeted incapacitation of the citizen by behavioral control. We expect that the consequences would be fatal in the long term, especially when considering the above-mentioned effect of undermining culture.

7.5   A Better Digital Society Is Possible

Despite fierce global competition, democracies would be wise not to cast the achievements of many centuries overboard. In contrast to other political regimes, Western democracies have the advantage that they have already learned to deal with pluralism and diversity. Now they just have to learn how to capitalize on them more.

In the future, those countries will lead that reach a healthy balance between business, government and citizens. This requires networked thinking and the establishment of an information, innovation, product and service "ecosystem." In order to work well, it is not only important to create opportunities for participation, but also to support diversity. Because there is no way to determine the best goal function: should we optimize the gross national product per capita or sustainability? Power or peace? Happiness or life expectancy? Often enough, what would have been better is only known after the fact. By allowing the pursuit of various different goals, a pluralistic society is better able to cope with the range of unexpected challenges to come.

Centralized, top-down control is a solution of the past, which is only suitable for systems of low complexity. Therefore, federal systems and majority decisions are the solutions of the present. With economic and cultural evolution, social complexity will continue to rise. Therefore, the solution for the future is collective intelligence. This means that citizen science, crowdsourcing and online discussion platforms are eminently important new approaches to making more knowledge, ideas and resources available.

Collective intelligence requires a high degree of diversity. This is, however, being reduced by today's personalized information systems, which reinforce trends.

Fig. 7.2 At the digital crossroads. Our society is at a crossroads: If ever more powerful algorithms would be controlled by a few decision-makers and reduce our self-determination, we would fall back in a Feudalism 2.0, as important historical achievements would be lost. Now, however, we have the chance to choose the path to digital democracy or democracy 2.0, which would benefit us all (see also https://vimeo.com/147442522)

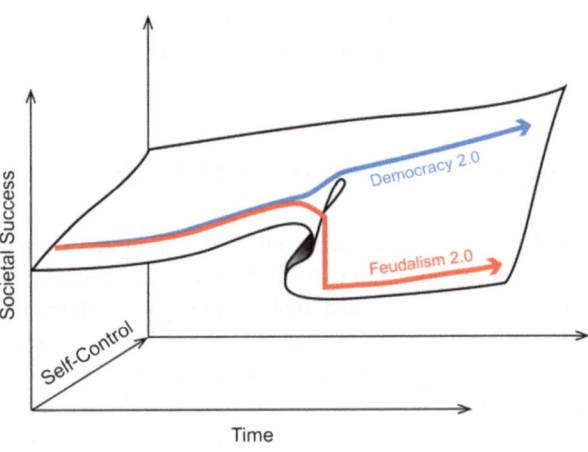

Sociodiversity is as important as biodiversity. It fuels not only collective intelligence and innovation, but also resilience—the ability of our society to cope with unexpected shocks. Reducing sociodiversity often also reduces the functionality and performance of an economy and society. This is the reason why totalitarian regimes often end up in conflict with their neighbors. Typical long-term consequences are political instability and war, as have occurred time and again throughout history. Pluralism and participation are therefore not to be seen primarily as concessions to citizens, but as functional prerequisites for thriving, complex, modern societies.

In summary, it can be said that we are now at a crossroads (see Fig. 7.2). Big data, artificial intelligence, cybernetics and behavioral economics are shaping our society—for better or worse. If such widespread technologies are not compatible with our society's core values, sooner or later they will cause extensive damage. They could lead to an automated society with totalitarian features. In the worst case, a centralized artificial intelligence would control what we know, what we think and how we act. We are at the historic moment, where we have to decide on the right path—a path that allows us all to benefit from the digital revolution. Therefore, we urge to adhere to the following fundamental principles:

1. to increasingly decentralize the function of information systems;
2. to support informational self-determination and participation;
3. to improve transparency in order to achieve greater trust;
4. to reduce the distortion and pollution of information;
5. to enable user-controlled information filters;
6. to support social and economic diversity;
7. to improve interoperability and collaborative opportunities;
8. to create digital assistants and coordination tools;
9. to support collective intelligence, and
10. to promote responsible behavior of citizens in the digital world through digital literacy and enlightenment.

Following this digital agenda we would all benefit from the fruits of the digital revolution: the economy, government and citizens alike. What are we waiting for?

7.6 A Strategy for the Digital Age

Big data and artificial intelligence are undoubtedly important innovations. They have an enormous potential to catalyze economic value and social progress, from personalized healthcare to sustainable cities. It is totally unacceptable, however, to use these technologies to incapacitate the citizen. Big nudging and citizen scores abuse centrally collected personal data for behavioral control in ways that are totalitarian in nature. This is not only incompatible with human rights and democratic principles, but also inappropriate to manage modern, innovative societies. In order to solve the genuine problems of the world, far better approaches in the fields of information and risk management are required. The research area of responsible innovation and the initiative "Data for Humanity" (see "Big Data for the benefit of society and humanity") provide guidance as to how big data and artificial intelligence should be used for the benefit of society.

What can we do now? First, even in these times of digital revolution, the basic rights of citizens should be protected, as they are a fundamental prerequisite of a modern functional, democratic society. This requires the creation of a new social contract, based on trust and cooperation, which sees citizens and customers not as obstacles or resources to be exploited, but as partners. For this, the state would have to provide an appropriate regulatory framework, which ensures that technologies are designed and used in ways that are compatible with democracy. This would have to guarantee informational self-determination, not only theoretically, but also practically, because it is a precondition for us to lead our lives in a self-determined and responsible manner.

There should also be a right to get a copy of personal data collected about us. It should be regulated by law that this information must be automatically sent, in a standardized format, to a personal data store, through which individuals could manage the use of their data (potentially supported by particular AI-based digital assistants). To ensure greater privacy and to prevent discrimination, the unauthorised use of data would have to be punishable by law. Individuals would then be able to decide who can use their information, for what purpose and for how long. Furthermore, appropriate measures should be taken to ensure that data is securely stored and exchanged.

Sophisticated reputation systems considering multiple criteria could help to increase the quality of information on which our decisions are based. If data filters and recommendation and search algorithms would be selectable and configurable by the user, we could look at problems from multiple perspectives, and we would be less prone to manipulation by distorted information.

In addition, we need an efficient complaints procedure for citizens, as well as effective sanctions for violations of the rules. Finally, in order to create sufficient transparency and trust, leading scientific institutions should act as trustees of the data and algorithms that currently evade democratic control. This would also require an appropriate code of conduct that, at the very least, would have to be followed by anyone with access to sensitive data and algorithms—a kind of Hippocratic Oath for IT professionals.

Furthermore, we would require a digital agenda to lay the foundation for new jobs and the future of the digital society. Every year we invest billions in the agricultural sector and public infrastructure, schools and universities—to the benefit of industry and the service sector.

Which public systems do we therefore need to ensure that the digital society becomes a success? First, completely new educational concepts are needed. This should be more focused on critical thinking, creativity, inventiveness and entrepreneurship than on creating standardised workers (whose tasks, in the future, will be done by robots and computer algorithms). Education should also provide an understanding of the responsible and critical use of digital technologies, because citizens must be aware of how the digital world is intertwined with the physical one. In order to effectively and responsibly exercise their rights, citizens must have an understanding of these technologies, but also of what uses are illegitimate. This is why there is all the more need for science, industry, politics, and educational institutions to make this knowledge widely available.

Secondly, a participatory platform is needed that makes it easier for people to become self-employed, set up their own projects, find collaboration partners, market products and services worldwide, manage resources and pay tax and social security contributions (a kind of sharing economy for all). To complement this, towns and even villages could set up centers for the emerging digital communities (such as fab labs), where ideas can be jointly developed and tested for free. Thanks to the open and innovative approach found in these centers, massive, collaborative innovation could be promoted.

Particular kinds of competitions could provide additional incentives for innovation, help increase public visibility and generate momentum for a participatory digital society. They could be particularly useful in mobilising civil society to ensure local contributions to global problems solving (for example, by means of "Climate Olympics"). For instance, platforms aiming to coordinate scarce resources could help unleash the huge potential of the circular and sharing economy, which is still largely untapped.

With the commitment to an open data strategy, governments and industry would increasingly make data available for science and public use, to create suitable conditions for an efficient information and innovation ecosystem that keeps pace with the challenges of our world. This could be encouraged by tax cuts, in the same way as they were granted in some countries for the use of environmentally friendly technologies.

Thirdly, building a "digital nervous system," run by the citizens, could open up new opportunities of the Internet of Things for everyone and provide real-time data measurements available to all. If we want to use resources in a more sustainable way and slow down climate change, we need to measure the positive and negative side effects of our interactions with others and our environment. By using appropriate feedback loops, systems could be influenced in such a way that they achieve the desired outcomes by means of self-organization.

For this to succeed we would need various incentive and exchange systems, available to all economic, political and social innovators. This could create entirely new markets and, therefore, also the basis for new prosperity. Unleashing the virtually unlimited potential of the digital economy would be greatly promoted by a pluralistic financial system (for example, functionally differentiated currencies) and new regulations for the compensation for inventions.

To better cope with the complexity and diversity of our future world and to turn it into an advantage, we will require personal digital assistants. These digital assistants will also benefit from developments in the field of artificial intelligence. In the future it can be expected that numerous networks combining human and artificial intelligence will be flexibly built and reconfigured, as needed. However, in order for us to retain control of our lives, these networks should be controlled in a distributed way. In particular, one would also have to be able to log in and log out as desired.

7.7 Democratic Platforms

A "Wikipedia of Cultures" could eventually help to coordinate various activities in a highly diverse world and to make them compatible with each other. It would make the mostly implicit success principles of the world's cultures explicit, so that they could be combined in new ways. A "Cultural Genome Project" like this would also be a kind of peace project, because it would raise public awareness for the value of sociocultural diversity. Global companies have long known that culturally diverse and multidisciplinary teams are more successful than homogeneous ones. However, the framework needed to efficiently collate knowledge and ideas from lots of people in order to create collective intelligence is still missing in many places. To change this, the provision of online deliberation platforms would be highly useful. They could also create the framework needed to realize an upgraded, digital democracy, with greater participatory opportunities for citizens. This is important, because many of the problems facing the world today can only be managed with contributions from civil society.

> **Information Box 1: Spotlight on China: Is This What the Future of Society Looks Like?**
> How would behavioural and social control impact our lives? The concept of a Citizen Score, which is now being implemented in China, gives an idea. There, all citizens are rated on a one-dimensional ranking scale. Everything they do gives plus or minus points. This is not only aimed at mass surveillance. The

(continued)

Information Box 1 (continued)

score depends on an individual's clicks on the Internet and their politically-correct conduct or not, and it determines their credit terms, their access to certain jobs, and travel visas. Therefore, the Citizen Score is about behavioural and social control. Even the behaviour of friends and acquaintances affects this score, i.e. the principle of clan liability is also applied: everyone becomes both a guardian of virtue and a kind of snooping informant, at the same time; unorthodox thinkers are isolated. Were similar principles to spread in democratic countries, it would be ultimately irrelevant whether it was the state or influential companies that set the rules. In both cases, the pillars of democracy would be directly threatened:

- The tracking and measuring of all activities that leave digital traces would create a "naked" citizen, whose human dignity and privacy would progressively be degraded.
- Decisions would no longer be free, because a wrong choice from the perspective of the government or company defining the criteria of the points system would have negative consequences. The autonomy of the individual would, in principle, be abolished.
- Each small mistake would be punished and no one would be unsuspicious. The principle of the presumption of innocence would become obsolete. Predictive Policing could even lead to punishment for violations that have not happened, but are merely expected to occur.
- As the underlying algorithms cannot operate completely free of error, the principle of fairness and justice would be replaced by a new kind of arbitrariness, against which people would barely be able to defend themselves.
- If individual goals were externally set, the possibility of individual self-development would be eliminated and, thereby, democratic pluralism, too.
- Local culture and social norms would no longer be the basis of appropriate, situation-dependent behaviour.
- The control of society with a one-dimensional goal function would lead to more conflicts and, therefore, to a loss of security. One would have to expect serious instability, as we have seen it in our financial system.

Such a control of society would turn away from self-responsible citizens to individuals as underlings, leading to a Feudalism 2.0. This is diametrically opposed to democratic values. It is therefore time for an Enlightenment 2.0, which would feed into a Democracy 2.0, based on digital self-determination. This requires democratic technologies: information systems, which are compatible with democratic principles—otherwise they will destroy our society.

Information Box 2: "Big Nudging"—Ill-designed for Problem Solving
*He who has large amounts of data can manipulate people in subtle ways. But even benevolent decision-makers may do more wrong than right, says **Dirk Helbing**.*

Proponents of Nudging argue that people do not take optimal decisions and it is, therefore, necessary to help them. This school of thinking is known as paternalism. However, Nudging does not choose the way of informing and persuading people. It rather exploits psychological weaknesses in order to bring us to certain behaviours, i.e. we are tricked. The scientific approach underlying this approach is called "behaviorism", which is actually long out of date.

Decades ago, Burrhus Frederic Skinner conditioned rats, pigeons and dogs by rewards and punishments (for example, by feeding them or applying painful electric shocks). Today one tries to condition people in similar ways. Instead of in a Skinner box, we are living in a "filter bubble": with personalized information our thinking is being steered. With personalized prices, we may be even punished or rewarded, for example, for (un)desired clicks on the Internet. The combination of Nudging with Big Data has therefore led to a new form of Nudging that we may call "Big Nudging". The increasing amount of personal information about us, which is often collected without our consent, reveals what we think, how we feel and how we can be manipulated. This insider information is exploited to manipulate us to make choices that we would otherwise not make, to buy some overpriced products or those that we do not need, or perhaps to give our vote to a certain political party.

However, Big Nudging is not suitable to solve many of our problems. This is particularly true for the complexity-related challenges of our world. Although already 90 countries use Nudging, it has not reduced our societal problems—on the contrary. Global warming is progressing. World peace is fragile, and terrorism is on the rise. Cybercrime explodes, and also the economic and debt crisis is not solved in many countries.

There is also no solution to the inefficiency of financial markets, as Nudging guru Richard Thaler recently admitted. In his view, if the state would control financial markets, this would rather aggravate the problem. But why should one then control our society in a top-down way, which is even more complex than a financial market? Society is not a machine, and complex systems cannot be steered like a car. This can be understood by discussing another complex system: our bodies. To cure diseases, one needs to take the right medicine at the right time in the right dose. Many treatments also have serious side and interaction effects. The same, of course, is expected to apply to social interventions by Big Nudging. Often is not clear in advance what would be good or bad for society. 60% of the scientific results in

(continued)

Information Box 2 (continued)

psychology are not reproducible. Therefore, chances are to cause more harm than good by Big Nudging.

Furthermore, there is no measure, which is good for all people. For example, in recent decades, we have seen food advisories changing all the time. Many people also suffer from food intolerances, which can even be fatal. Mass screenings for certain kinds of cancer and other diseases are now being viewed quite critically, because the side effects of wrong diagnoses often outweigh the benefits. Therefore, if one decided to use Big Nudging, a solid scientific basis, transparency, ethical evaluation and democratic control would be really crucial. The measures taken would have to guarantee statistically significant improvements, and the side effects would have to be acceptable. Users should be made aware of them (in analogy to a medical leaflet), and the treated persons would have to have the last word.

In addition, applying one and the same measure to the entire population would not be good. But far too little is known to take appropriate individual measures. Not only is it important for society to apply different treatments in order to maintain diversity, but correlations (regarding what measure to take in what particular context) matter as well. For the functioning of society it is essential that people apply different roles, which are fitting to the respective situation they are in. Big Nudging is far from being able to deliver this.

Current Big-Data-based personalization rather creates new problems such as discrimination. For instance, if we make health insurance rates dependent on certain diets, then Jews, Muslims and Christians, women and men will have to pay different rates. Thus, a bunch of new problems is arising.

Richard Thaler is, therefore, not getting tired to emphasize that Nudging should only be used in beneficial ways. As a prime example, how to use Nudging, he mentions a GPS-based route guidance system. This, however, is turned on and off by the user. The user also specifies the respective goal. The digital assistant then offers several alternatives, between which the user can freely choose. After that, the digital assistant supports the user as good as it can in reaching the goal and in making better decisions. This would certainly be the right approach to improve people's behaviours, but today the spirit of Big Nudging is quite different from this.

Information Box 3: Digital Self-Determination by Means of a "Right to a Copy"

Europe must guarantee citizens a right to a digital copy of all data about them (Right to a Copy), says **Ernst Hafen.** *A first step towards data democracy*

(continued)

Information Box 3 (continued)

would be to establish cooperative banks for personal data that are owned by the citizens rather than by corporate shareholders.

Medicine can profit from health data. However, access to personal data must be controlled the persons (the data subjects) themselves. The "Right to a Copy" forms the basis for such a control.

In Europe, we like to point out that we live in free, democratic societies. We have almost unconsciously become dependent on multinational data firms, however, whose free services we pay for with our own data. Personal data— which is now sometimes referred to as a "new asset class" or the oil of the twenty first Century—is greatly sought after. However, thus far nobody has managed to extract the maximum use from personal data because it lies in many different data sets. Google and Facebook may know more about our health than our doctor, but even these firms cannot collate all of our data, because they rightly do not have access to our patient files, shopping receipts, or information about our genomic make-up. In contrast to other assets, data can be copied with almost no associated cost. Every person should have the right to obtain a copy of all their personal data. In this way, they can control the use and aggregation of their data and decide themselves whether to give access to friends, another doctor, or the scientific community.

The emergence of mobile health sensors and apps means that patients can contribute significant medical insights. By recording their bodily health on their smartphones, such as medical indicators and the side effects of medications, they supply important data which make it possible to observe how treatments are applied, evaluate health technologies, and conduct evidence-based medicine in general. It is also a moral obligation to give citizens access to copies of their data and allow them to take part in medical research, because it will save lives and make health care more affordable.

European countries should copper-fasten the digital self-determination of their citizens by enshrining the "Right to a Copy" in their constitutions, as has been proposed in Switzerland. In this way, citizens can use their data to play an active role in the global data economy. If they can store copies of their data in non-profit, citizen-controlled, cooperative institutions, a large portion of the economic value of personal data could be returned to society. The cooperative institutions would act as trustees in managing the data of their members. This would result in the democratization of the market for personal data and the end of digital dependence.

Information Box 4: Democratic Digital Society

In order to deal with future technology in a responsible way, it is necessary that each one of us can participate in the decision-making process, argues **Bruno S. Frey** *from the University of Basel.*

How can responsible innovation be promoted effectively? Appeals to the public have little, if any, effect if the *institutions* or rules shaping human interactions are not designed to incentivize and enable people to meet these requests.

Several types of institutions should be considered. Most importantly, society must be *decentralized*, following the principle of subsidiarity. Three dimensions matter.

- *Spatial* decentralization consists in vibrant federalism. The provinces, regions and communes must be given sufficient autonomy. To a large extent, they must be able to set their own tax rates and govern their own public expenditure.
- *Functional* decentralization according to area of public expenditure (for example education, health, environment, water provision, traffic, culture etc.) is also desirable. This concept has been developed through the proposal of FOCJ, or "Functional, Overlapping and Competing Jurisdictions".
- *Political* decentralization relating to the division of power between the executive (government), legislative (parliament) and the courts. Public media and academia should be additional pillars.

These types of decentralization will continue to be of major importance in the digital society of the future.

In addition, citizens must have the opportunity to directly participate in decision-making on particular issues by means of popular referenda. In the discourse prior to such a referendum, all relevant arguments should be brought forward and stated in an organized fashion. The various proposals about how to solve a particular problem should be compared and narrowed down to those which seem to be most promising, and integrated insomuch as possible during a mediation process. Finally, a referendum needs to take place, which serves to identify the most viable solution for the local conditions (viable in the sense that it enjoys a diverse range of support in the electorate).

Nowadays, on-line deliberation tools can efficiently support such processes. This makes it possible to consider a larger and more diverse range of ideas and knowledge, harnessing "collective intelligence" to produce better policy proposals.

Another way to implement the 10 proposals would be to create new, unorthodox institutions. For example, it could be made compulsory for every official body to take on an *"advocatus diaboli"*. This lateral thinker would be tasked with developing counter-arguments and alternatives to each

(continued)

Information Box 4 (continued)
proposal. This would reduce the tendency to think along the lines of "political correctness" and unconventional approaches to the problem would also be considered.

Another unorthodox measure would be to choose among the alternatives considered reasonable during the discourse process using *random decision-making mechanisms*. Such an approach increases the chance that unconventional and generally disregarded proposals and ideas would be integrated into the digital society of the future.

Information Box 5: Democratic Technologies and Responsible Innovation
*When technology determines how we see the world, there is a threat of misuse and deception. Thus, innovation must reflect our values, argues **Jeroen van den Hoven**.*

Germany was recently rocked by an industrial scandal of global proportions. The revelations led to the resignation of the CEO of one of the largest car manufacturers, a grave loss of consumer confidence, a dramatic slump in share price and economic damage for the entire car industry. There was even talk of severe damage to the "Made in Germany" brand. The compensation payments will be in the range of billions of Euro.

The background to the scandal was a situation whereby VW and other car manufacturers used manipulative software which could detect the conditions under which the environmental compliance of a vehicle was tested. The software algorithm altered the behavior of the engine so that it emitted fewer pollutant exhaust fumes under test conditions than in normal circumstances. In this way, it cheated the test procedure. The full reduction of emissions occurred only during the tests, but not in normal use.

In the twenty first Century, we urgently need to address the question of how we can implement ethical standards technologically.

Similarly, algorithms, computer code, software, models and data will increasingly determine what we see in the digital society, and what are choices are with regard to health insurance, finance and politics. This brings new risks for the economy and society. In particular, there is a danger of deception.

Thus, it is important to understand that our values are embodied in the things we create. Otherwise, the technological design of the future will determine the shape of our society ("code is law"). If these values are self-

(continued)

Information Box 5 (continued)

serving, discriminatory or contrary to the ideals of freedom and personal privacy, this will damage our society. Thus, in the twenty first Century we must urgently address the question of how we can implement ethical standards technologically. The challenge calls for us to "design for value".

If we lack the motivation to develop the technological tools, science and institutions necessary to align the digital world with our shared values, the future looks very bleak. Thankfully, the European Union has invested in an extensive research and development program for responsible innovation. Furthermore, the EU countries which passed the Lund and Rome Declarations emphasized that innovation needs to be carried out responsibly. Among other things, this means that innovation should be directed at developing intelligent solutions to societal problems, which can harmonize values such as efficiency, security and sustainability. Genuine innovation does not involve deceiving people into believing that their cars are sustainable and efficient. Genuine innovation means creating technologies that can actually satisfy these requirements.

Information Box 6: Digital Risk Literacy

Rather than letting intelligent technology diminish our brainpower, we should learn to better control it, says **Gerd Gigerenzer**—*beginning in childhood.*

The digital revolution provides an impressive array of possibilities: thousands of apps, the Internet of Things, and almost permanent connectivity to the world. But in the excitement, one thing is easily forgotten: innovative technology needs competent users who can control it rather than be controlled by it.

Three examples:

One of my doctoral students sits at his computer and appears to be engrossed in writing his dissertation. At the same time his e-mail inbox is open, all day long. He is in fact waiting to be interrupted. It's easy to recognize how many interruptions he had in the course of the day by looking at the flow of his writing.

An American student writes text messages while driving:

"When a text comes in, I just have to look, no matter what. Fortunately, my phone shows me the text as a pop up at first. . . so I don't have to do too much looking while I'm driving." If, at the speed of 50 miles per hour, she takes only 2 seconds to glance at her cell phone, she's just driven 48 yards "blind". That young woman is risking a car accident. Her smart phone has taken control of her behavior—as is the case for the 20–30% of Germans who also text while driving.

(continued)

Information Box 6 (continued)

During the parliamentary elections in India in 2014, the largest democratic election in the world with over 800 million potential voters, there were three main candidates: N. Modi, A. Kejriwal, and R. Ghandi. In a study, undecided voters could find out more information about these candidates using an Internet search engine. However, the participants did not know that the web pages had been manipulated: For one group, more positive items about Modi popped up on the first page and negative ones later on. The other groups experienced the same for the other candidates. This and similar manipulative procedures are common practice on the Internet. It is estimated that for candidates who appear on the first page thanks to such manipulation, the number of votes they receive from undecided voters increases by 20% points.

In each of these cases, human behavior is controlled by digital technology. Losing control is nothing new, but the digital revolution has increased the possibility of that happening.

What can we do? There are three competing visions. One is techno-paternalism, which replaces (flawed) human judgment with algorithms. The distracted doctoral student could continue readings his emails and use thesis-writing software; all he would need to do is input key information on the topic. Such algorithms would solve the annoying problem of plagiarism scandals by making them an everyday occurrence.

Although still in the domain of science fiction, human judgment is already being replaced by computer programs in many areas. The BabyConnect app, for instance, tracks the daily development of infants—height, weight, number of times it was nursed, how often its diapers were changed, and much more—while newer apps compare the baby with other users' children in a real-time database. For parents, their baby becomes a data vector, and normal discrepancies often cause unnecessary concern.

The second vision is known as "nudging". Rather than letting the algorithm do all the work, people are steered into a particular direction, often without being aware of it. The experiment on the elections in India is an example of that. We know that the first page of Google search results receives about 90% of all clicks, and half of these are the first two results. This knowledge about human behavior is taken advantage of by manipulating the order of results so that the positive ones about a particular candidate or a particular commercial product appear on the first page. In countries such as Germany, where web searches are dominated by one search engine (Google), this leads to endless possibilities to sway voters. Like techno-paternalism, nudging takes over the helm.

But there is a third possibility. My vision is risk literacy, where people are equipped with the competencies to control media rather than be controlled by

(continued)

Information Box 6 (continued)

it. In general, risk literacy concerns informed ways of dealing with risk-related areas such as health, money, and modern technologies. Digital risk literacy means being able to take advantage of digital technologies without becoming dependent on or manipulated by them. That is not as hard as it sounds. My doctoral student has since learned to switch on his email account only three times a day, morning, noon, and evening, so that he can work on his dissertation without constant interruption.

Learning digital self-control needs to begin as a child, at school and also from the example set by parents. Some paternalists may scoff at the idea, stating that humans lack the intelligence and self-discipline to ever become risk literate. But centuries ago the same was said about learning to read and write—which a majority of people in industrial countries can now do. In the same way, people can learn to deal with risks more sensibly. To achieve this, we need to radically rethink strategies and invest in people rather than replace or manipulate them with intelligent technologies. In the twenty first century, we need less paternalism and nudging and more informed, critical, and risk-savvy citizens. It's time to snatch away the remote control from technology and take our lives into our own hands.

Information Box 7: Ethics—Big Data for the Common Good and for Humanity

The power of data can be used for good and bad purposes. **Roberto Zicari** *and* **Andrej Zwitter** *have formulated five principles of Big Data Ethics.*

In recent times there have been a growing number of voices—from tech visionaries like Elon Musk (Tesla Motors), to Bill Gates (Microsoft) and Steve Wozniak (Apple)—warning of the dangers of artificial intelligence (AI). A petition against automated weapon systems was signed by 200,000 people and an open letter recently published by MIT calls for a new, inclusive approach to the coming digital society.

We must realize that big data, like any other tool, can be used for good and bad purposes. In this sense, the decision by the European Court of Justice against the Safe Harbour Agreement on human rights grounds is understandable.

States, international organizations and private actors now employ big data in a variety of spheres. It is important that all those who profit from big data are aware of their moral responsibility. For this reason, the Data for Humanity Initiative was established, with the goal of disseminating an ethical code of

(continued)

Information Box 7 (continued)
conduct for big data use. This initiative advances five fundamental ethical principles for big data users:

1. *"Do no harm"*. The digital footprint that everyone now leaves behind exposes individuals, social groups and society as a whole to a certain degree of transparency and vulnerability. Those who have access to the insights afforded by big data must not harm third parties.
2. *Ensure that data is used in such a way that the results will foster the peaceful coexistence of humanity.* The selection of content and access to data influences the world view of a society. Peaceful coexistence is only possible if data scientists are aware of their responsibility to provide even and unbiased access to data.
3. *Use data to help people in need.* In addition to being economically beneficial, innovation in the sphere of big data could also create additional social value. In the age of global connectivity, it is now possible to create innovative big data tools which could help to support people in need.
4. *Use data to protect nature and reduce pollution of the environment.* One of the biggest achievements of big data analysis is the development of efficient processes and synergy effects. Big data can only offer a sustainable economic and social future if such methods are also used to create and maintain a healthy and stable natural environment.
5. *Use data to eliminate discrimination and intolerance and to create a fair system of social coexistence.* Social media has created a strengthened social network. This can only lead to long-term global stability if it is built on the principles of fairness, equality and justice.

To conclude, we would also like to draw attention to how interesting new possibilities afforded by big data could lead to a better future: "As more data become less costly and technology breaks barriers to acquisition and analysis, the opportunity to deliver actionable information for civic purposes grows. This might be termed the 'common good' challenge for big data." (Jake Porway, DataKind). In the end, it is important to understand the turn to big data as an opportunity to do good and as a hope for a better future.

Information Box 8: Measuring, Analyzing, Optimizing—When Intelligent Machines Take Over Societal Control
In the digital age, machines steer everyday life to a considerable extent already. We should, therefore, think twice before we share our personal data, says expert **Yvonne Hofstetter.**

(continued)

Information Box 8 (continued)

If Norbert Wiener (1894–1964) had experienced the digital era, for him it would have been the land of plenty. "Cybernetics is the science of information and control, regardless of whether the target of control is a machine or a living organism", the founder of Cybernetics once explained in Hannover, Germany in 1960. In history, the world never produced such amount of data and information as it does today.

Cybernetics, a science asserting ubiquitous importance, makes a strong claim: "Everything can be controlled." During the twentieth century, both the US armed forces and the Soviet Union applied Cybernetics to control their arms' race. The NATO had deployed so-called C3I systems (Command, Control, Communication and Information), a term for military infrastructure that leans linguistically to Wiener's book on *Cybernetics: Or Control and Communication in the Animal and the Machine,* published in 1948. Control refers to the control of machines as well as of individuals or entire social systems like military alliances, financial markets or, pointing to the twenty first century, even the electorate. Its major premise: keeping the world under surveillance to collect data. Connecting people and things to the *Internet of Everything* is a perfect to way to obtain the required mass data as input to cybernetic control strategies.

With Cybernetics, Wiener proposed a new scientific concept: the closed-loop feedback. Feedback—e.g. the *Likes* we give, the online comments we make—is a major concept of digitization, too. Does that mean digitization is the most perfect implementation of Cybernetics? When we use smart devices, we are creating a ceaseless data stream disclosing our intentions, geo position or social environment. While we communicate more thoughtlessly than ever online, in the background, an ecosystem of artificial intelligence is evolving. Today, artificial intelligence is the sole technology being able to profile us and draw conclusions about our future behavior.

An automated control strategy, usually a learning machine, analyzes our actual situation and then computes a stimulus that should draw us closer to a more desirable "optimal" state. Increasingly, such controllers govern our daily lives. As digital assistants they help us making decisions in the vast ocean of optionality and intimidating uncertainty. Even Google Search is a control strategy. When typing a keyword, a user reveals his intentions. The Google search engine, in turn, will not just present a list with best hits, but a link list that embodies the highest (financial) value rather for the company than for the user. Doing it that way, i.e. listing corporate offerings at the very top of the search results, Google controls the user's next clicks. This, the European Union argues, is a misuse.

But is there any way out? Yes, if we disconnected from the cybernetic loop. Just stop responding to a digital stimulus. Cybernetics will fail, if the

(continued)

Information Box 8 (continued)

controllable counterpart steps out of the loop. Yet, we are free to owe a response to a digital controller. However, as digitization further escalates, soon we may have no more choice. Hence, we are called on to fight for our freedom rights—afresh during the digital era and in particular at the rise of intelligent machines.

For Norbert Wiener (1894–1964), the digital era would be a paradise. "Cybernetics is the science of information and control, regardless of whether a machine or a living organism is being controlled", the founder of cybernetics once said in Hanover, Germany in 1960.

Cybernetics, a science which claims ubiquitous importance makes a strong promise: "Everything is controllable." During the twentieth century, both the US armed forces and the Soviet Union applied cybernetics to control the arms' race. NATO had deployed so-called C3I systems (Command, Control, Communication and Information), a term for military infrastructure that linguistically leans on Wiener's book entitled *Cybernetics: Or Control and Communication in the Animal and the Machine* published in 1948. Control refers to the control of machines as well as of individuals or entire societal systems such as military alliances, NATO and the Warsaw Pact. Its basic requirements are: Integrating, collecting data and communicating. Connecting people and things to the *Internet of Everything* is a perfect way to obtain the required data as input of cybernetic control strategies.

With cybernetics, a new scientific concept was proposed: the closed-loop feedback. Feedback—such as the *likes* we give or the online comments we make—is another major concept related to digitization. Does this mean that digitization is the most perfect implementation of cybernetics? When we use smart devices, we create an endless data stream disclosing our intentions, geolocation or social environment. While we communicate more thoughtlessly than ever online, in the background, an artificial intelligence (AI) ecosystem is evolving. Today, AI is the sole technology able to profile us and draw conclusions about our future behavior.

An automated control strategy, usually a learning machine, analyses our current state and computes a stimulus that should draw us closer to a more desirable "optimal" state. Increasingly, such controllers govern our daily lives. Such digital assistants help us to make decisions among the vast ocean of options and intimidating uncertainty. Even Google Search is a control strategy. When typing a keyword, a user reveals his intentions. The Google search engine, in turn, presents not only a list of the best hits, but also a list of links sorted according to their (financial) value to the company, rather than to the user. By listing corporate offerings at the very top of the search results, Google controls the user's next clicks. That is a misuse of Google's monopoly, the European Union argues.

(continued)

Information Box 8 (continued)

But is there any way out? Yes, if we disconnect from the cybernetic loop and simply stop responding to the digital stimulus. Cybernetics will fail, if the controllable counterpart steps out of the loop. We should remain discreet and frugal with our data, even if it is difficult. However, as digitization further escalates, soon there may be no more choices left. Hence, we are called on to fight once again for our freedom in the digital era, particularly against the rise of intelligent machines.

Further Reading

ACLU: Orwellian Citizen Score, China's credit score system, is a warning for Americans, http://www.computerworld.com/article/2990203/security/aclu-orwellian-citizen-score-chinas-credit-score-system-is-a-warning-for-americans.html

Big data, meet Big Brother: China invents the digital totalitarian state. The worrying implications of its social-credit project. The Economist (December 17, 2016)

Frey, B.S., Gallus, J.: Beneficial and exploitative nudges. In: Economic analysis of law in European legal scholarship. Springer, Heidelberg (2015)

Gigerenzer, G.: On the supposed evidence for libertarian paternalism. Review of Philosophy and Psychology. 6(3), S. 361–S. 383 (2015)

Grassegger, H., Krogerus, M: Ich habe nur gezeigt, dass es die Bombe gibt [I have only shown the bomb exists]. Das Magazin (3. Dezember 2016). https://www.dasmagazin.ch/2016/12/03/ich-habe-nur-gezeigt-dass-es-die-bombe-gibt/

Hafen, E., Kossmann, D., Brand, A.: Health data cooperatives—citizen empowerment. Methods of Information in Medicine. 53(2), S. 82–S. 86 (2014)

Harris, S.: The Social Laboratory, Foreign Policy (29 July 2014). http://foreignpolicy.com/2014/07/29/the-social-laboratory/

Helbing, D.: The automation of society is next: How to survive the digital revolution. CreateSpace, 2015

Helbing, D.: Thinking ahead—Essays on big data, digital revolution, and participatory market society. Springer, Cham (2015)

Helbing, D., Pournaras, E.: Build digital democracy. Nature. 527, S. 33–S. 34 (2015)

van den Hoven, J., Vermaas, P.E., van den Poel, I.: Handbook of ethics, values and technological design. Springer, Dordrecht (2015)

Volodymyr, M., Kavukcuoglu, K., Silver, D., et al.: Human-level control through deep reinforcement learning. Nature. 518, S. 529–S. 533 (2015)

Tong, V.J.C.: Predicting how people think and behave, International Innovation, http://www.internationalinnovation.com/predicting-how-people-think-and-behave/

Zicari, R., Zwitter, A.: Data for humanity: An open letter. Frankfurt Big Data Lab, 13.07.2015. Zwitter, A.: Big Data Ethics. In: Big Data & Society 1(2), 2014

Dirk Helbing is Professor of Computational Social Science at the Department of Humanities, Social and Political Sciences and affiliate professor at the Department of Computer Science at ETH Zurich. His recent studies discuss globally networked risks. At Delft University of Technology he directs the PhD programme "Engineering Social Technologies for a Responsible Digital Future." He is also an elected member of the German Academy of Sciences "Leopoldina" and the World Academy of Art and Science.

Bruno S. Frey is an economist and Visiting Professor at the University of Basel, where he directs the Center for Research in Economics and Well-Being (CREW). He is also Research Director of the Center for Research in Economics, Management and the Arts (CREMA) in Zurich.

Gerd Gigerenzer is Director at the Max Planck Institute for Human Development in Berlin and the Harding Center for Risk Literacy, founded in Berlin in 2009. He is a member of the Berlin-Brandenburg Academy of Sciences and the German Academy of Sciences "Leopoldina". His research interests include risk competence and risk communication, as well as decision-making under uncertainty and time pressure.

Ernst Hafen is Professor at the Institute of Molecular Systems Biology at ETH Zurich and also its former President. In 2012, he founded the initiative "Data and Health." The initiative's intention is to strengthen citizens' digital self-determination at a political and economic level, as well as to encourage the establishment of organised cooperative databases for personal data.

Michael Hagner is Professor of Science Studies at ETH Zurich. His research interests include the relationship between science and democracy, the history of cybernetics and the impact of digital culture on academic publishing.

Yvonne Hofstetter is a lawyer and AI expert. The analysis of large amounts of data and data fusion systems are her specialities. She is the Managing Director of Teramark Technologies GmbH. The company develops digital control systems based on artificial intelligence, for, among other purposes, the optimisation of urban supply chains and algorithmic currency risk management.

Jeroen van den Hoven is University Professor and Professor of Ethics and Technology at Delft University of Technology, as well as founding Editor in Chief of the journal of Ethics and Information Technology. He was founding Chairman of the Dutch Research Council program on Responsible Innovation and chaired an Expert Group Responsible Research and Innovation of the European Commission. He is member of the Expert Group on Ethics of the European Data Protection Supervisor.

Roberto V. Zicari is Professor for Databases and Information Systems at the Goethe University Frankfurt and Big Data expert. His interests also include entrepreneurship and innovation. He is the founder of the Frankfurt Big Data Lab at the Goethe University and the editor of the Operational Database Management Systems (ODBMS.org) portal. He is also a Visiting Professor at the Center for Entrepreneurship and Technology of the Department of Industrial Engineering and Operations Research at the University of California at Berkeley.

Andrej Zwitter is Professor of International Relations and Ethics at the University of Groningen, in the Netherlands, and Honorary Senior Research Fellow at Liverpool Hope University, U.K. He is the co-founder of the International Network Observatory for Big Data and Global Strategy. His research interests include international political theory, emergency and martial law, humanitarian aid policy, as well as the impact of Big Data on international politics and ethics.

Chapter 8
Digital Fascism Rising?

Dirk Helbing

Can we still stop a world of technological totalitarianism?

Any claim that we humans are (already) contending with a new form of—this time digital—fascism will immediately be discredited as overblown.

No wonder: There are very powerful business forces who each make tens of billions of dollars a year by singing the sweet song of how our existence as individuals, as well as democracy in general, is enhanced by the conveniences of Big Data and Artificial Intelligence.

However, the real issue is whether democracy, such as we know it, can survive Big Data and Artificial Intelligence. We have long entered a world rife with new kinds of behavioral manipulation.

Whether we like to admit it or not, today's secret services and Big Data companies possess much more data about us than were needed to run totalitarian states in the past. It is unlikely that such power will not be misused at some point in time.

This article by Dirk Helbing was first published on October 20, 2017, in The Globalist under the URL https://www.theglobalist.com/fascism-big-data-artificial-intelligence-surveillance-democracy/ (reprinted with permission)
An English-language video with the author exploring the themes laid out here in more detail, can be found under the URL https://www.youtube.com/watch?v=-hgnuYZ__ms

D. Helbing (✉)
ETH Zurich, Zürich, Switzerland

TU Delft, Delft, Netherlands

Complexity Science Hub, Vienna, Austria
e-mail: dhelbing@ethz.ch

8.1   Features of Digital Societies

Any doubter about my core claim—that we are running into conditions of a new, digital kind of fascism—should consider the following list of features of many modern digital societies:

- mass surveillance,
- unethical experiments with humans,
- social engineering,
- forced conformity ("Gleichschaltung"),
- propaganda and censorship,
- "benevolent" dictatorship,
- (predictive) policing,
- different valuation of people,
- relativity of human rights,
- and, it seems, possibly even euthanasia for the expected times of crisis in our unsustainable world.

That list exhibits all the core aims that fascists in the past were dreaming of.

We humans have been sweet talked collectively to the point that we don't even realize this new, digitally empowered kind of totalitarianism.

8.2   Saving Democracy

Even worse, in the past, the conditions for fascism were usually just found in select countries. In the digital era, the new fascism has reached global dimensions.

If we want to save democracy, freedom and human dignity, an emergency operation is inevitable. Often heard arguments to justify digital fascism—such as the need to fight terrorism, cyber threats and climate change—have been skilfully used to undermine our privacy, our rights and democracy itself.

The emergence of mass surveillance after 9/11, enabled by the Patriot Act in the United States and other laws, has led to the incremental erosion of liberties and human rights. Since the Snowden revelations, we know that there is mass surveillance of billions of people around the world.

But most people still have no idea how pervasive it is, and how it may influence their lives in the future. Billions of dollars have been spent on mass surveillance tools of secret services to hack our computers, smartphones, smart TVs and smart cars.

The estimated amount of data collected about us every day ranges from millions of numbers to Gigabytes of data. As a result, we have ended up with the digital tools for a data-driven, AI-based so-called "benevolent" dictatorship, where big businesses and the state determine "what is best for us."

Citizens are being targeted, their data collected and consolidated. This is used to create a near complete profile of each person, their nature, habits and preferences. Each profile can contain thousands of specifiers.

8.3 Manipulating Behavior

As if that weren't bad enough in itself, these digital doubles can be used to make thousands of computer experiments with our virtual self to find out how our thinking and behavior can be manipulated.

More specifically, our personal data is being applied to customize information such that it will influence our attention, emotions, opinions, decisions and behaviours—often subconsciously—by a technique called big nudging or neuro-marketing.

This ranges from steering our consumption behaviour to manipulating voting behaviour in elections.

In the wrong hands, the misuse of surveillance-based personal data will have catastrophic consequences for us as individuals and for society as a whole. In an explicitly or implicitly totalitarian state, this kind of information could be used to predict and identify those people who don't agree with certain government policies and sanction them even before they can exercise their democratic rights.

The British secret service, for example, runs a program called Karma Police,[1] which shows where our societies are heading. This Citizen Score, which is currently also tested in China, may be used to run an entirely new kind of autocratic society, or even police state.

According to plans, the Citizen Score would determine the level of access to facilities, products and services. We would be scored or penalized according to our behaviours. Reading critical news or having the "wrong" kinds of social ties, for example, would get you minus points.

8.4 Countering Digital Totalitarianism

To counter this danger of a digital totalitarian state, we must at a minimum ensure:

- a democratic framework of use for powerful cyberinfrastructures,
- scientific use by interdisciplinary teams, considering multiple perspectives,
- ethical use considering human rights and human dignity,
- transparency,
- cyber-security (by decentralization etc.),

---

[1]https://en.wikipedia.org/wiki/Karma_Police_%28surveillance_program%29, https://theintercept. com/2015/09/25/gchq-radio-porn-spies-track-web-users-online-identities/

- informational self-determination (e.g., with a Personal Data Store)

The door is wide open for global fascism to take hold, unless we take action now. The longer we wait, the harder it will be to fight back, as more—and more intrusive—data are collected every single day, and powerful algorithms are used to predict and—increasingly control—our behavior in its totality.

That is why all of us are well advised to pay attention to well-founded concerns regarding the rise of a technological totalitarianism that—this time—is unfolding on a global scale since the data collectors really know no boundaries.

8.5   Concerning Global Events

Events around the world are quite concerning. The recent German election has just seen the unprecedented rise of a right-wing party, in part promoted by voter targeting and social bots.

Countries such as Spain, Hungary, Poland and Turkey are clearly on the path to more authoritarianism.

Political developments in France, the UK, United States, Japan, Austria, Switzerland and the Netherlands deserve attention, too. Some of their politicians have started to question human rights. Other people have even openly sympathized with Hitler.

Politics—that means, all of us—must act. It cannot be ignored anymore that civilization, as we built it after World War II, is at stake. Our societies are in a danger of derailing.

It is important to take time to think about the future we really want to live in and to leave old kinds of thinking behind. We need a real public discourse and a positive vision of our future.

Moreover, the old powers must allow change to happen. It's time to re-invent society, but fascism should be left behind once and for all!

8.6   Takeaways

- Intrusive data collection practices make a new, digital fascism—via the data-driven steering of masses of humans—a real possibility.
- The issue is whether democracy, such as we know it, can survive Big Data and Artificial Intelligence.
- Today's secret services and Big Data companies possess much more data about us than were needed to run totalitarian states in the past.
- We humans have been sweet talked collectively to the point that we don't even realize the new, digitally empowered kind of totalitarianism.
- In the past, conditions for fascism were usually just found in select countries. In the digital era, fascism has reached global dimensions.

Chapter 9
The Birth of a Digital God

Dirk Helbing

Will a superintelligent system in the near future know everything and tell us how to live?

It is finally happening! At the annual meeting of the Swiss Civil Society Association on November 11, Professor Hans Ulrich Gumbrecht gave a memorable speech—a "mass," as some listeners thought. It was not just about trying to create a super-intelligent system with consciousness. No, the goal was now to create a God-like being with superhuman knowledge and abilities to guide our human destiny. However, there is the risk that this God might turn against humanity, he continued, even though it was man-made. The statement that this should free us from Biblical sin was even more surprising.

Gumbrecht is not the first one to raise the subject of Artificial Intelligence (AI) as God. Just recently, the Guardian, under the title "Deus Ex Machina," announced that ex-Google collaborator Levandowski wanted to register Artificial Intelligence as religion.[1] Shortly later, Google announced its latest triumph. They had succeeded in building an AI system that learned to win the strategy game "Go" by itself—so well in fact that it could beat the world champion. At the same time, it was suggested that one had now found an approach that would sooner or later solve all the problems of humanity, including those that surpassed our intellectual capacities.

This article by Dirk Helbing appeared first in German on November 25, 2017, in the newspaper "Schweiz am Sonntag" under the URL https://www.aargauerzeitung.ch/kommentare-aaz/geburt-eines-digitalen-gottes-131939639

[1] https://www.theguardian.com/technology/2017/sep/28/artificial-intelligence-god-anthony-levandowski

D. Helbing (✉)
ETH Zurich, Zürich, Switzerland

TU Delft, Delft, Netherlands

Complexity Science Hub, Vienna, Austria
e-mail: dhelbing@ethz.ch

© Springer International Publishing AG, part of Springer Nature 2019
D. Helbing (ed.), *Towards Digital Enlightenment*,
https://doi.org/10.1007/978-3-319-90869-4_9

Just a few days later, Spiegel Online wrote: "God does not need any teachers."[2]
Already in 2013, I discussed the opportunities and risks of the information age in an
article entitled "Google as God?"[3] Furthermore, in 2015, the Digital Manifesto
asked: "Let us suppose there was a super-intelligent machine with God-like knowledge and superhuman abilities: would we follow its instructions?"[4]

Some readers found the question ridiculous at that time. Not anymore! Because
search engines and intelligence services know almost everything about us. We have
been living in a Big Brother world already for some time. George Orwell's dystopian
novel "1984," written in 1948, was meant as a warning. But more and more often we
get the feeling the bestseller was actually used as an instruction manual.

Today's data-driven world has two main principles: "Data is the new oil" and
"Knowledge is power." Little by little, and almost unnoticed, this has created a
fundamentally new society. There is a new currency, "data," which replaces classical
money. There is a new economic system: the "attention economy," where our
attention is sold by auctions in split seconds. In addition, the companies of "surveillance capitalism" are measuring our behavior, our personality and our lives in ever
more detail. In times of free services, we have become a product ourselves. Last but
not least, the principle "code is law" has established a new legal system, which
bypasses our parliament.

Are we in danger of losing our liberties, human rights and participation step by
step, almost imperceptibly? Are we giving up on things that are important to us, just
because we fear terrorism, climate change, and cybercrime? Are self-determined
citizens in a danger to be turned into remotely controlled subjects?

In fact, this isn't just phantasy! China is already testing a Citizen Score,[5] i.e. every
citizen is rated, has a certain number of points. Minus points will punish those who
do not pay for their loan immediately, cross the street during a red light, have the
"wrong" friends or neighbors, or reads critical news. The Citizen Score then
determines the job opportunities, loan conditions, access to services, and mobility
restrictions. Great Britain seems to go even a step further. It measures its citizens
including the videos they watch and the music they hear. The system is called
"Karma Police."[6] So, will it punish thought crimes, you may ask? Or is "Karma
Police" a kind of "Judgment Day" waiting to come down on us any time?

Do we have to accept this? Computers make better decisions, it is often said. In
fact, computers have been the better chess players for years. In many areas they are
better workers. They don't get tired, do not complain, do not go on vacation, and do

---

[2]http://www.spiegel.de/wissenschaft/technik/kuenstliche-intelligenz-gott-braucht-keine-
lehrmeister-kolumne-a-1175130.html

[3]https://www.nzz.ch/google-als-gott-1.18049950

[4]http://www.spektrum.de/thema/das-digital-manifest/1375924, English translation: https://www.
scientificamerican.com/article/will-democracy-survive-big-data-and-artificial-intelligence/

[5]https://www.economist.com/news/briefing/21711902-worrying-implications-its-social-credit-pro
ject-china-invents-digital-totalitarian

[6]https://theintercept.com/2015/09/25/gchq-radio-porn-spies-track-web-users-online-identities/

not have to pay taxes and contributions to social security. Soon they will be better drivers. They diagnose cancer better than physicians and answer questions better than people—at least those that have already an answer.

When will robots become our judge and hangman? When will they start to "fix the overpopulation problem"? (Autonomous killer robots with face recognition probably exist already or could at least exist soon—see the recent movies on slaughterbots and robot swarms.[7]) When will robots replace us? Not just our work… A newspaper article recently suggested that the descendants of humans will be machines.[8] In other words, humanity will be replaced by robots. Is this really our human destiny? Should we build a future for robots or for humans? Isn't it time to wake up from the transhumanist dream?[9]

Back to the initial question: Is Google creating a digital God? With its Loon project, the company at least tries to be omnipresent. With its search engine, language assistants and measurement sensors in our rooms, Google wants to be omniscient. While the company is not yet omnipotent, it is at least answering 95% of our questions, and with personalized information, Google is increasingly steering our thinking and actions. Furthermore, the Calico project is also trying to make people immortal. Therefore, in an overpopulated world, would Google be the judge over life and death?

Whatever, someone recently suggested an AI God would soon write a new Bible.[10] So would he (or she) set the rules we would have to live by? Do we soon have to worship an AI algorithm and submit ourselves to it? No question, some already seem to dream of a digital God who will guide our human destiny. What for some is the invention of God through human ingenuity, however, must be the ultimate blasphemy for Christians—in some sense the rise of the Antichrist.

Whatever one may think about all this, the phrase "knowledge is power" has certainly blown some people's minds. Google, IBM and Facebook are said to be working on a new operating system for society.[11] Democracy is defamed as outdated technology.[12] They want to engineer paradise on Earth—a smarter planet where

---

[7]https://www.youtube.com/watch?v=9CO6M2HsoIA,   https://www.youtube.com/watch?v=CGAk5gRD-t0

[8]https://www.nzz.ch/feuilleton/unsere-nachfahren-werden-maschinen-sein-ld.1322780

[9]https://www.nzz.ch/meinung/kommentare/die-gefaehrliche-utopie-der-selbstoptimierung-wider-den-transhumanismus-ld.1301315,   http://privacysurgeon.org/blog/wp-content/uploads/2017/07/Human-manifesto_26_short-1.pdf

[10]https://venturebeat.com/2017/10/02/an-ai-god-will-emerge-by-2042-and-write-its-own-bible-will-you-worship-it/

[11]http://www.faz.net/aktuell/feuilleton/medien/google-gruendet-in-den-usa-government-innovaton-lab-13852715.html,   https://www.pcworld.com/article/3031137/forget-trump-and-clinton-ibms-watson-is-running-for-president.html,   https://www.theguardian.com/technology/2017/feb/17/facebook-ceo-mark-zuckerberg-rule-world-president,   http://theconversation.com/if-facebook-ruled-the-world-mark-zuckerbergs-vision-of-a-digital-future-73459

[12]Hencken, Randolph. 2014. In: Mikrogesellschaften. Hat die Demokratie ausgedient? Capriccio. Video, veröffentlicht am 15.5.2014. Autor: Joachim Gaertner. München: Bayerischer Rundfunk.

everything will be automated. So far, however, the plan did not really work out.[13] The world's cities with the highest quality of life are located everywhere, but in the leading IT nations. And even in the Silicon Valley, the heart of the digital revolution, and other IT hotspots, experts start to worry. . .

Elon Musk, for example, fears that Artificial Intelligence could become the greatest threat to humanity. Even Bill Gates had to admit that he was in the camp of those who were worried about superintelligence. The famous physicist Stephen Hawking warned that humans would not be able to compete with the development of Artificial Intelligence. Apple co-founder Steve Wozniak agreed: "Computers are going to take over from humans, no question," he said, but: "Will we be the gods? Will be the family pets? Or will be ants that get stepped on? I don't know. . ."[14] Jürgen Schmidhuber, German AI pioneer, believes to know—from a robot's perspective, we will be like cats.[15]

Of course, the worry that technology could turn against us is already old. Besides George Orwell's "1984" and "Animal Farm," Aldous Huxley's "Brave New World" warned us of the danger of rising totalitarianism. Suddenly people also remember "The Machine Stops" by Edward Morgan Forster in 1909 (!). More recent books are Dave Egger's "The Circle," "Homo Deus" by Yuval Noah Harari and Joel Cachelin's "Internet God." If you like science fiction, you might love "QualityLand" by Marc-Uwe Kling or "iGod" by Willemijn Dicke.

A question, which not only science fiction lovers should ask, is: What future do we want to live in? Never before have we had a better chance to build a world of our liking. But for this we have to take the future into our hands. It's high time to overcome our self-imposed digital immaturity. To free ourselves from the digital shackles, digital literacy and enlightenment are needed. So far, we are living in a market-conform democracy, where the markets are driven by technology. Instead, we should build an economy that serves to reach the goals of people and society. Technology should be a means of achieving this. This requires a fundamental redesign of our monetary, financial and economic system based on the principle of value-sensitive design. In "The Globalist," I have recently outlined how this could be done.[16] Maybe you have your own ideas of how to use Big Data and Artificial Intelligence. But in any case, a better future is possible! Let's demand this better future! Let's co-create it! What are we waiting for?

---

[13]https://www.wiltonpark.org.uk/wp-content/uploads/WP1449-Report.pdf

[14]https://www.computerworld.com/article/2901679/steve-wozniak-on-ai-will-we-be-pets-or-mere-ants-to-be-squashed-our-robot-overlords.html

[15]http://www.faz.net/aktuell/feuilleton/debatten/ueberwindung-des-menschen-durch-selbstlernende-maschinen-15309705.html

[16]https://www.theglobalist.com/author/dirk-helbing/

Chapter 10
To the Elites of the World: Time to Act

Dirk Helbing

Faced with climate change, financial, economic and spending crisis, mass migration, terrorism, wars and cyber threats, it appears we are very close to global emergency. Given this state of affairs, we are running out of time to fix the problems of our planet. Here, we present what should be decided during the UN General Assembly on September 23, 2017 and a reflexive preamble.

We acknowledge your efforts to improve the quality of life. However, these efforts have also caused a further increase in the consumption of resources and energy.

It appears that this is now driving our planet to the edge: Climate change affects the global water system, agriculture and the basis of the lives of billions of people. It causes environmental disasters, mass migration and armed conflicts. Moreover, it is estimated to threaten about one-sixth of all species on our planet.

Nevertheless, global disaster is not inevitable—if we re-organize the world in a suitable way, as discussed below.

The lives of billions of people are at risk. It is the moral duty of politicians, religious, cultural, scientific and business leaders—in short: the elite—to avert likely disasters, humanitarian crises and ethical dilemmas as much as possible. This

---

A spiced-up version of this chapter by Dirk Helbing was published as FuturICT Blog on July 31, 2017, under the URL http://futurict.blogspot.ch/2017/07/to-elites-of-world.html. Most of the blog appeared as a series of seven opinion pieces in The Globalist in August 2017 (reprinted here with permission). The first piece is accessible under the URL https://www.theglobalist.com/population-environment-technology-society-climate-change-disaster/; the other links can be found right there.

D. Helbing (✉)
ETH Zurich, Zürich, Switzerland

TU Delft, Delft, Netherlands

Complexity Science Hub, Vienna, Austria
e-mail: dhelbing@ethz.ch

© Springer International Publishing AG, part of Springer Nature 2019   107
D. Helbing (ed.), *Towards Digital Enlightenment*,
https://doi.org/10.1007/978-3-319-90869-4_10

requires bringing about the necessary changes of society on the way in a timely manner.

With the aim to "save the planet," many have urged the world community to reduce carbon emissions drastically by 2030 and almost completely by the end of the century. However, given that the world population has grown roughly proportional to the global oil and gas consumption, such a drop would largely reduce the carrying capacity of the Earth for people—unless the reduction in carbon-based energy can be replaced by renewable energy in a timely manner.

New solutions are needed not only for heating and transportation, but also for the chemical industry, because the production of plastic and fertilizer currently depends on oil. Altogether, a radical re-organization of major parts of our economy appears to be urgently necessary.

Even though philanthropy and engagement in responsible innovation have increased, this urgent transformation has not taken place to the required extent. To a considerable degree, this is because those who have "vested interests" in the old system have often obstructed change.

However, "vested interests" are no excuse for inaction or delays. Property and power imply responsibility. If this responsibility is not adequately exercised, power lacks legitimacy.

If people have to pay with their lives for "vested interests", these interests clearly undermine the very basis of societies.

Human dignity, which underpins many fundamental values and human rights, is the imperative that all individual, political and economic action should be oriented at. It is the key value and central pillar of many modern societies and, according to many constitutions, must be *actively* protected by all means.

10.1   A Final Call to Action

If humanity wants to bring a positive future or even a "Golden Age of Prosperity and Peace" on the way, we need to dramatically reform our basic societal institutions, e.g. the present financial and monetary system, our economy and society.

Even though it seems that the current organizational principles of our world have served us well for a long time, they are now often failing to deliver the right solutions early enough.

Within the current framework, time and again we got trapped in suboptimal solutions to complex coordination games, "tragedies of the commons" and problems of collective inaction.

In our highly networked cyber-physical world, "linear thinking" (the assumption that effects are proportional to their causes) and the ethics of small-group, face-to-face interactions in relatively simple settings are often leading us astray.

Fundamental change is inevitable. It seems that what needs to take center stage now is not how much money or power someone can accumulate, but how much he or she is benefitting others and the world. Apparently, our societies have largely lost track of this basic guiding principle.

Claiming that our problem is overpopulation of the planet reveals lack of imagination.

By now it is obvious that all traditional problem-solving approaches have failed to work.

Also, the attempt to revive historical forms of societal organization, empowered by Big Data and Artificial Intelligence, does not seem to work, as the recent experiences in various countries with technocratic Smart Cities approaches have shown.[1]

However, if innovation within the current system is not sufficient, the system itself has to be reinvented and changed.

It seems paradoxical that—in times of an abundance of data and the best technology ever—centralized control attempts failed to boost our most advanced economies and societies to a new level of satisfaction and prosperity, sustainability and resilience.

The reason for this lies in the complexity of hyper-connected systems, in which processing power cannot keep up with data volumes and those cannot keep up with the combinatorial increase in complexity.

Such networked systems often behave in unexpected and counter-intuitive ways: Rather than the intended effects, one will frequently find side effects, feedback effects and cascading effects.

Given these circumstances, centralized control attempts perform often poorly. Even the most powerful artificial intelligence systems will not be able to manage the overly complex and often quickly changing systems of our globalized world well enough. As a consequence, a new, decentralized control paradigm is needed, which implies the need for modular designs, diverse solutions, and participatory opportunities.[2]

Therefore, we need new ways of participatory decision-making as well as new designs of the monetary, financial and economic system. In the new framework we propose, co-creation, co-ordination, co-evolution and collective intelligence are the main underlying success principles.

---

[1]Report on the Wilton Park Event "Disrupting cities through technology", see https://www.wiltonpark.org.uk/wp-content/uploads/WP1449-Report.pdf

[2]D. Helbing, The Automation of Society Is Next: How to Survive the Digital Revolution (CreateSpace, 2015).

10.2 Rethinking the World Economy: From Push to Pull

Obviously, the world's supply chains must be organized in a completely different way. What we need is a combined circular and sharing economy, as many have pointed out.

Presently, however, because it's often cheaper, we have many "linear" supply chains. In these, fresh resources are used to produce large numbers of products for the sake of economies of scale, which are then sold ("pushed") to as many customers as possible using massive marketing campaigns. The customers will then consume the products—and eventually throw them away.

Supply chains must hence be organized in a better way.

Rather than today's "push economy", we need an economy driven by demands, i.e. a "pull economy". The world's resources would be enough for everyone, if we reused and shared them.

Today, inhabitants of the industrialized world produce about 50 tons of waste in a lifetime. This includes several cars, computers, smartphones, a lot of furniture and other things that would probably be enough for, say, five people.

In principle, our planet could offer a higher quality of life for more people with less resources.

Reusing and recycling these resources would need renewable energy. Such energy has gained increasing market shares in many places, but the focus has been on *big* solutions—such as power stations that would produce energy for hundreds of thousands of people, as this implies the most attractive business models.

There is an increasing amount of evidence that more energy-efficient or environment-friendly solutions have often been suppressed by established industries in attempts to maintain their "cash cows".

In order to bring a better future for mankind in view, we need to reinvent about half of our economy within a few decades. This is possible, but the process needs to start now.

It is obvious that the current governance system of the world has failed to deliver the needed solutions on time. Therefore, *a new social contract should be urgently made.*

The United Nations General Assembly on September 23, 2017 is the right place for this to take place. Decisions in accordance with these proposals should be implemented in the 7 years following this date.

10.3 A New Social Contract

It is time to work out a new social contract that allows everyone to lead a proper life and determine it to a larger extent.

Rather than advancing their own benefits, elites can—and must—be formidable agents of change for good. Therefore, they should (be able to) live up to their responsibilities for the future of our planet and coming generations in proportion to their wealth, power and positions.

10.4   Reform of the United Nations

The United Nations have been a useful instrument to progress world affairs in a number of areas. However, it has failed to produce world peace, to eliminate the production and proliferation of arms, and to reduce the level of inequality in the world.

The United Nations Agenda 2030 shows that there are a lot of urgent matters still on the to do list.

Most people concur with the idea that we should have a global organization of world affairs, and that national egoism should be largely overcome.

In contrast to the current concept of a "world government", however, we need a more decentralized, participatory and diverse approach, which leaves freedom to experiment with new solutions.

Therefore, regions should play a more important role.[3] They could form global cooperation networks to address shared problems more effectively.

For example, they could regularly perform City Olympics, i.e. engage in friendly competitions to innovate, find, and implement the best, energy-saving, environmental-friendly, socially responsible, sustainable and resilient solutions. To foster open innovation, these solutions shall be shared with the world under open source and Creative Commons licenses.

Lobbying by industrial and other interests should be replaced by a transparent mechanism of policy and decision making. Therefore, industry representatives should sit in the World Council suggested below, while in the future, traditional lobbying should be abolished by law (as it constitutes an intransparent form of getting political influence).

In order to make sure that decision-making will be based on facts, science should be represented in the envisaged World Council too. The presence of citizen representatives should, furthermore, ensure that the interests of normal people are represented as well.

The following illustration represents the suggested composition of the proposed World Council:

---

[3]This new form of globalisation may be called "glocalisation": the principle would be to think global, but act local (and diverse), experiment, and learn from each other.

A PROPOSAL: REPLACE THE UN WITH THIS

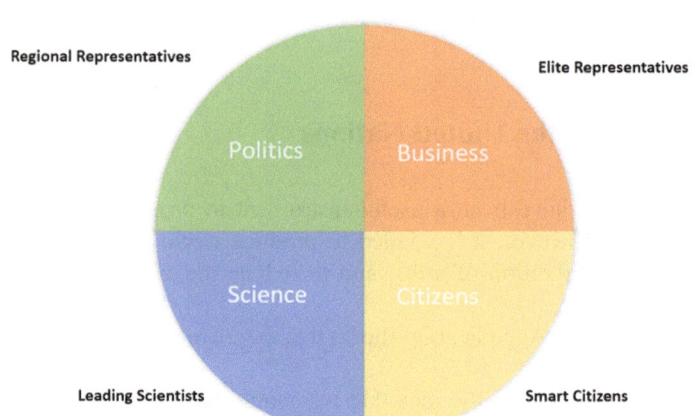

- Each of the four sections of the World Council would be of equal size (e.g., 1000 representatives each). The Council should aim to achieve the best possible balance over world regions and interests.
- There should be no veto right. Instead, binding votes should require a "grand majority" of two third. If this grand majority is not reached, one should offer choices to enable locally fitting solutions and some degree of diversity, e.g. by creating a "best of list of solutions". If an urgent vote must be taken and the choice of a single solution is inevitable, the proposed solution getting the highest number of votes should be implemented, but the solution should be temporary in nature, carefully evaluated, and taken back, if necessary.
- The first and main duty of members of the Council is to serve the interest of the world, and their activities must be fully transparent (in particularly sensitive matters, activities may be recorded and disclosed with a 10 year delay). Decisions should be taken on the basis of individual insights, not formal or informal memberships of political parties or interest groups.
- Members of the Council will have to completely disclose their property, sources of income, formal or informal memberships, special opportunities, and anything that might compromise independence or create a conflict of interest.
- The business sector may devise their own rules to select their representatives, but the rules should be approved by the other three council parts.
- Citizen representatives would be chosen in each region based on an open competition transmitted by public media, in which willing participants would demonstrate their knowledge and commitment to the public interest.
- Scientific representatives should be internationally leading experts, who are economically independent and cover the scope of fields and disciplines in a balanced way. Their research must be funded in full by public sources. It is inacceptable for these members to pursue (or have pursued in the past 7 years) research on behalf of companies or foundations, as these may have a special

agenda and bias the scope of research or amount of resources invested in certain questions, approaches, or solutions.

- Representatives from the four sectors would have to be completely disentangled). Family or other ties, interest groups, political parties or other parties are strictly discouraged in the interest of representative, unbiased decision-making. The attempt to undermine independent decision-making will be sanctioned by exclusion from the World Council.
- The World Council will establish and run "democratic capitalism" and a multidimensional, socio-ecological finance or incentive system as described below. Its members may get a certain percentage of new value created by it, i.e. payoffs shall be performance-based. Taxes (money to create public goods) shall also be directly derived from this new monetary and financial system.
- The Council's may be led by a 24 person group of people, composed of six elected representatives of each section, which would be coordinated by a chair person. The decisions of this Steering Group would be preliminary and would have to approved by the World Council. Otherwise they will run out by the end of the next World Council meeting.

10.5   Upgrade of Today's Capitalism

As today's form of capitalism is not compatible with our social and cultural value system, capitalism is sooner or later damaging the foundations of societies and the values they are built on.

Therefore, capitalism must be upgraded in a way that is compatible with societal and cultural values and with the fairness principle to provide equal opportunities.

Since a change of the world's carrying capacity by 1% effectively decides over the lives and deaths of about 80 million people, it is unacceptable that innovations are obstructed or restricted to those that are compatible with current business models.

The survival of billions of people will depend on our ability to drastically increase innovation rates and to generate more pluralistic innovations.

In other words, *it is morally imperative to enable mass innovation, as neither venture capitalism nor philanthropy nor other standard means of supporting innovation were sufficient* to solve the existential problems of our planet. We are quite far from having made it a place where *all* people can live in dignity and unfold their talents.

For this and a number of other reasons, the monetary system needs urgent reform. The current system is not fair and creates serious distortions. It further promotes inequality, which creates political instability.

In fact, it tends to undermine the very basis of democracy and other institutions. The current monetary system implies existential and political dependence, which constrains individual and collective development.

Therefore, in the future, everyone shall have equal opportunities to unfold their talents and engagement. This shall also include the right to benefit from money generation.

From a legal point of view, everyone should be equal, and hence this should also apply to money creation.

10.6 Democratic Capitalism, Crowd Funding for All

Benefitting from money creation can no longer be the privilege of a few private persons and banks.

Moreover, in times of digital, ecological and societal transformation, everyone should be able to experiment and discover new solutions.

Therefore, everyone shall soon get a universal investment premium.[4] This money will not be provided from taxes, but by money generation.

The overall amount of money in circulation shall be kept at bay by a negative interest rate.[5] The underlying idea is to take money out of the system that is not being used, because it is desirable for the economy and society that money is being invested.

The investment premium shall not be kept or spent by the person who receives it.[6] It should rather be invested into the best ideas and projects engaging for social and environmental affairs, new technologies, improved neighbourhoods, etc.

This "crowd funding for all" may be realized by a new kind of money. Its height shall make sure that the better half (or at least a third) of proposed projects can be realized.

People will be able to earn an additional income by winning projects and contributing to their realization, such that there is a mechanism encouraging innovation and engagement.

Project results realized with these investment premiums shall become open source and Creative Commons after a two-year time period,[7] such that combinatorial

---

[4]Potentially together with a basic income. Both would be perfectly compatible with most constitutions, as long as currently existing property would not be taken away.

[5]This would not necessarily have to be realized through inflation as we know it today. A decaying value could also be produced by giving each amount of money a specific "creation date" and defining a value that depends on its "age". It would now be possible to create electronic coins and paper bills that have such a feature built in.

[6]Nor should it be invested into family members—with the exception of publicly approved cases, where these people depend on the help of others for health reasons; investments into the same person or project shall also be limited to restrict undesirable kickback deals.

[7]We have currently a similar time-limited protection for patents, so this approach would be compatible with already existing legal principles.

innovation and a participatory information, innovation, product and service system can emerge and thrive.

Note that everyone would benefit from this approach, and it is not expected that this will be to the disadvantage of large companies.[8]

The above described measures are intended to *boost the massive, pluralistic innovation that will now be needed to solve the world's existential problems collectively as soon as possible.*

Within just a few years, half of the economy will have to be reinvented to make it sustainable and create new jobs in the wake of automation that it is now driven by Artificial Intelligence and Robotics.

This requires existential security, experimental opportunities and access to innovative and productive means.

If the above reforms are made, a redistribution of property from the rich to the poor may be avoidable—otherwise it will be inevitable.

In any case, property that is not actively used by the owner(s) for more than a year, should be made accessible for use through a sharing economy platform that provides fair access to anyone who is qualified for use.

Private property that is not publicly registered shall be administered by the region where the property is located.

Likewise, services and products that have been discontinued, patents or business models that are not being used, shall become Creative Commons within 2 years' time.

10.7 Reforms of the Financial System

Due to the implications of the new wave of automation on the job market and the need to create a resilient[9] and sustainable world, we have to re-invent about half of our economy.

We have just a few decades to build an ecological, digital economy that consumes much less coal, gas, oil, water and other resources.

At the same time, it is important to be able to distinguish innovations that hurt nature or other people from those that don't, and to promote the latter.

Sustainability, resilience and ethical alignment could be reached with a new, differentiated incentive system—a "socio-ecological finance system" that is efficient, innovation-friendly and democratic at the same time.

---

[8]This ecosystem approach is inspired by the rain forest, where it's the dropping of leaves after some time, which creates humus and, thereby, the basis of abundance and diversity. Note that the biggest trees grow in rain forests.

[9]Resilience, i.e. the ability to quickly recover from shocks and successfully respond to unexpected developments, requires, in particular, a sufficient degree of decentralization and diversity.

How can this be achieved? *The financial system is essentially a coordination mechanism, which decides who receives how much of what resource at what price.*

But there could be a myriad of better coordination systems. Instead of managing society with a complicated tax system with a 1–2 year delay, the Internet of Things will soon enable a real-time management of complex systems, namely by real-world measurements and real-time feedback.

This can be set up in such a way that the values of society are built into the system ("values by design").

With the Internet of Things, the effects of our actions, including our "externalities", can now be measured at low cost: noise, stress, carbon dioxide (CO_2), other emissions, waste, etc.

The same applies to desired outcomes such as job creation, social cooperation, education, health and the reuse of resources.

These would be attributed a price or value in the socio-ecological finance system, which would be determined in a participatory way based on a subsidiary organization of the world.[10]

With the addition of numerous new currencies,[11] existing alongside today's one-dimensional monetary system, one could increase the desired effects and activities and reduce unwanted ones. Social and ecological commitment would no longer be expensive—it would pay off.

With such a multi-dimensional feedback and incentive system, a circular economy would basically emerge by itself, driven by new market forces rather than regulation or a digital command economy.

Numerous regulations could be replaced by measurement processes and participatory (subsidiary) pricing processes.

Through a hierarchy of incentive systems, one could promote local commitment to achieve global goals. The economy would become resource-efficient and driven by people's needs.

Businesses and citizens could benefit alike. In the interest of digital democracy and collective intelligence, the socio-ecological financial system would be jointly managed by representatives of the economy, politics, science and the general public as described above in the section on the World Council.

In addition, the socio-ecological finance system could be designed in such a way that it would automatically generate taxes to pay for public goods and infrastructures.

---

[10]The subsidiarity principle assumes a multi-level organization and demands that something should be decided on the lowest level possible, if reasonably efficient, and on a higher level only if necessary. This principle allows for individually, locally, and culturally fitting solutions, which enable everyone to unfold their talents. "Economies of scale" may not be a sufficient reason to replace regional diversity by standardized products and solutions. Pluralism and diversity are the basis of innovation, societal resilience, and collective intelligence. Hence, they must be sufficiently protected from over-standardization and the control by too few companies, institutions, or people.

[11]To create the desired multi-dimensionality, exchange between the various currencies will have to be discouraged by introducing a suitable amount of "friction" (such as conversion taxes).

By means of the differentiated, multi-dimensional incentive system, one could manage complex systems much better, and even build self-organizing or self-regulating systems.

The externalities underlying this incentive system would be measured in a crowd-sourced way, using sensors in smartphones and the Internet of Things.

By sharing the measurements and making the data available to all, one could earn different kinds of money. Even without a redistribution of money and wealth, everyone could benefit, simply by organizing the use of resources much better.

10.8   Digital Upgrade of Democracy[12] ("Digital Democracy")

Around the world, many democracies have come about as the response to revolutions and wars. Therefore, their defining features reflect the lessons learned through history.

These features include human dignity and human rights, the respect of a private sphere (in the sense of protection from exposure or misuse and the right to be left alone), self-determination, pluralism and protection of minorities, checks and balances, the separation of powers, anonymous and equal votes, equal opportunities, transparency, fairness, legitimacy and justice.

Good education, enlightenment and empowerment of people in order to enable them to make constructive contributions to our collective future are important elements of modern societies, too.

Social media have recently increased participatory opportunities, but they also have some drawbacks. They have been criticized for promoting hate speech, filter bubbles and echo chambers, polarization and extremism, fake news and disinformation, as well as the manipulation of emotions, opinions, decisions and behavior.

Against this backdrop, it has been claimed that democracy and the wisdom of crowds do not work in the digital age. New, data-driven, technocratic ways of decision-making would be more efficient and should, hence, replace *democracy, which is claimed to be an "outdated technology"*.

However, we need social systems that are able to produce alternative, better and diverse solutions to the complex problems we are faced with.

In particular, what matters for the performance of economies and societies is that people can unfold their knowledge, ideas, talents and resources well. This requires a

---

[12]For a more detailed description of the relevant aspects see "How to make democracy work in the digital age" by Dirk Helbing and Stefan Klauser, http://www.huffingtonpost.com/entry/how-to-make-democracy-work-in-the-digital-age_us_57a2f488e4b0456cb7e17e0f; Why we need democracy 2.0 and capitalism 2.0 to survive, published in Jusletter IT, see https://papers.ssrn.com/soL3/papers.cfm?abstract_id=2769633

societal framework that is oriented at increasing co-creation opportunities for all and harnessing collective intelligence.

The creation of collective intelligence requires a good educational system, reliable, unbiased information, independent search of information and solutions, and diversity. Under such conditions, the combination of several solutions creates often a better solution to a complex problem than the single best solution.

Constructive forms of massive open online deliberation (MOOD) require new kinds of participatory platforms, which allow people affected by a problem to contribute arguments, ideas and concerns to the related debate.

These contributions would have to be organized in a logical, fact-based argument graph that works out the various perspectives on a complex problem and its various implications for diverse kinds of stakeholders.

Artificial Intelligence could help to organize the arguments, while experienced and trusted people should moderate the process in an unbiased manner.

Once the different arguments and perspectives are clear and possible solutions have been suggested, one should start a round table with key representatives of the different perspectives to work out integrated solutions in an innovative deliberation process.

A voting of the affected people on a "best of" list of integrated solutions should then decide, which of the integrated solutions fits the needs of the people in the respective region best. It should hence be implemented there.

10.9   Design and Operation of Information and Other Man-Made Systems

To ensure that man-made systems are aligned with the fundamental (constitutional) values of society, including human dignity, one needs to demand ethically aligned design ("design for values", "value-sensitive design"). Man-made systems should be also culturally adaptive and provide possibilities for users to customize them to their personal needs by providing options. It is furthermore needed

1. to increasingly decentralize the function of information systems;
2. to support informational self-determination and participation (e.g. through personal data stores and opt-out possibilities);
3. to improve transparency in order to achieve greater trust;
4. to reduce the biases and pollution of information;
5. to enable user-controlled information filters;
6. to support social and economic diversity;
7. to improve interoperability and collaborative opportunities;
8. to create digital assistants and coordination tools;
9. to support collective intelligence,
10. to promote responsible behaviour of citizens in the digital world through digital literacy and enlightenment,

11. to make algorithms and robots identifiable and ensure (a suitable system of individual or collective) accountability for information systems—or other products and services,[13]
12. to implement "kill switches" in order to be able to terminate systems that hurt people or violate their rights or undermine the constitution.

Experiments with humans (including decision-experiments such as A/B-testing, psychological conditioning, etc.) need to be sufficiently transparent, comply with ethical standards (including informed consent), and allow for opt-out.

Information displayed in print or on the Web or elsewhere should be categorized at least into the following different types:

- facts (objective contributions—links to original sources in peer-reviewed high-level science journals [or other highly credible sources] would have to be provided),
- opinions, claims, questions etc. (subjective contributions),
- advertisements (information provided to potentially create financial or other benefits),
- personalized [information] (information that has been customized to the respectively targeted recipients, using personal data).

Furthermore, it should be distinguishable whether the sender of the information is

- a person (displaying the real identity),
- anonymous,
- pseudonymous (displaying a made-up identity),
- an algorithm or bot.

Misclassifications should be suitably sanctioned. Filtering according to categories should be possible. It should also be possible to turn personalization on and off, and to determine the degree of personalization (how many personal parameters are being considered).

Furthermore, "war rooms" should be replaced by "peace rooms". As compared to a war room setting, a "peace room" is characterized by a number of additional features such as: a higher degree of transparency (to reduce possible flaws and increase trust), a democratic framework of operation (for legitimacy), the use by interdisciplinary teams meeting international scientific standards (to achieve the integration of the best knowledge available), the supervision by ethical experts (to ensure responsible use and innovation), a multi-stakeholder and multi-perspective approach (to find solutions that work for everyone—as much as this is possible), and, in order to increase problem solving capacity, participatory opportunities for civil society (by means of NGOs, citizen science, and/or crowd sourcing).

---

[13] Analogous to the principle "Parents are liable for their children".

10.10 Guiding Principles for a "Golden Age of Prosperity and Peace"

Science, technology and our inventions have allowed us to dominate the Earth and everything that lives on it. In the meantime, we have increasingly understood what the side effects and impacts of our man-made interventions are.

Therefore, it is now time for an ethics that makes us fit for our future.

So, what kind of fundamental values may be guiding us in the densely connected, digital society of the future?

1. **Respect:** Treat all forms of life respectfully; protect and promote their (mental, psychological and physical) well-being.
2. **Diversity and non-discrimination:** Support socio-economic diversity and pluralism (also by the ways in which Information and Communications Technologies are designed and operated). Counter discrimination and repression, prioritize rewards over punishment.
3. **Freedom:** Support the principle of informational self-determination; respect creative freedom (opportunities for individual development) and the freedom of non-intimidating expression.
4. **Participatory opportunities:** Enable self-determined decisions, offer participatory opportunities and a choice of good options. Ensure to properly balance the interests of all relevant (affected) stakeholders, particularly political and business interests, and those of citizens.
5. **Self-organization:** Create a framework to support flexible, decentralized, self-organized adaptation, e.g. by using suitable reputation systems.
6. **Responsibility:** Commit yourself to timely, responsible and sustainable actions (or omissions), by considering their externalities.
7. **Quality and awareness:** Commit yourself to honest, high-quality information and good practices and standards; support transparency and awareness.
8. **Fairness:** Reduce negative externalities that are directly or indirectly caused by your own decisions and actions, and fully compensate the disadvantaged parties.
9. **Protection:** Protect others from harm, damage, and exploitation; refrain from aggressive or war-like activities (including cybercrime, cyberwar, and misuse of information).
10. **Resilience:** Reduce the vulnerability of systems and increase their resilience (e.g. through decentralization, self-organization and diversity).
11. **Sustainability:** Promote sustainable systems and long-term societal benefits; increase systemic benefits.
12. **Compliance:** Engage in protecting and complying with these fundamental principles.

To summarize the above even more briefly, the most important rule is *to increase positive externalities, to reduce negative ones, and to be fair,* or to formulate it even shorter: *"Try not to harm others or nature!"*

Chapter 11
Why We Need Democracy 2.0 and Capitalism 2.0 to Survive

Dirk Helbing

The world is running into great trouble. The Anthropocene challenges (including climate change, impending resource shortages, demographic change, conflict, financial and economic crises) call for entirely new answers. As a result, we are now seeing the emergence of data-driven societies around the globe. Feudalism 2.0, fascism 2.0, communism 2.0, socialism 2.0, democracy 2.0 and capitalism 2.0 can now be built. What framework should we choose? What would be the implications?

In the recent past, we have seen companies and governments around the world collect increasing amounts of personal and other data. This has been justified with better customer experience and with security issues, e.g. the need to fight organized crime and international terrorism. However, statistical investigations do not confirm that mass surveillance is more efficient in identifying and preventing terrorism than conventional means.[1] In fact, all individuals who have recently committed terror

---

This article by Dirk Helbing was first published on 25 May 2016 under the same title in Jusletter IT (weblaw.ch) (reprinted with permission)

[1]The effectiveness of predictive policing: Lessons from a randomized controlled trial, Journalist's Resource (6 November 2014), http://journalistsresource.org/studies/government/criminal-justice/predictive-policingrandomized-controlled-trial (all internet sources have been last visited on 25 April 2016); Kai Biermann, Predictive Policing—Noch hat niemand bewiesen, dass Data Mining der Polizei hilft, ZEIT Online (29 March 2015), http://www.zeit.de/digital/datenschutz/2015-03/predictive-policing-software-polizei-precobs; Ellen Nakashima, NSA phone record collection does little to prevent terrorist attacks, group says, The Washington Post (12 January 2014), http://www.washingtonpost.com/world/national-security/nsa-phone-recordcollection-doeslittle-to-prevent-terrorist-attacks-group-says/2014/01/12/8aa860aa-77dd-11e3-8963-b4b654bcc9b2_story.

D. Helbing (✉)
ETH Zurich, Zürich, Switzerland

TU Delft, Delft, Netherlands

Complexity Science Hub, Vienna, Austria
e-mail: dhelbing@ethz.ch

© Springer International Publishing AG, part of Springer Nature 2019
D. Helbing (ed.), *Towards Digital Enlightenment*,
https://doi.org/10.1007/978-3-319-90869-4_11

attacks in Europe were known before to be potentially dangerous extremists.[2] Nevertheless, the attacks have not been prevented. Therefore, can mass surveillance ever serve the declared purpose? And if not, why is it so energetically pursued?

Recently, it has become increasingly clear that personal data about everyone of us is not only collected for the sake of personalized advertisements, services and products,[3] but also politically used in various countries. The keyword here is "Big Nudging", i.e. the combination of methods from behavioral economics ("nudging") and Big Data to manipulate the attention, opinions, decisions, and behaviors of people.[4] The more data is available about us, the easier it becomes to manipulate us subconsciously, i.e. without our awareness. It is suggested that this method would be used to make us behave in a healthier and environmentally friendly way, but the method has also been applied to manipulate other public opinions and elections.[5] Politics has started to notice the danger of this approach, which enables new kinds of propaganda and misinformation, while it also undermines social cohesion by means of the "filter bubble effect".[6] Furthermore, it reduces critical thinking, which makes people vulnerable to populist and extremist opinions.

---

html?hpid=z4;   see   also   http://web.archive.org/web/20150211220452/http://securitydata. newamerica.net/nsa/analysis

[2]Sascha Lobo, Die Mensch-Maschine: Tiefgreifendes, strukturelles, multiples Staatsversagen, Spiegel Online (30 March 2016), http://www.spiegel.de/netzwelt/netzpolitik/sascha-lobo-ueber-is-terror-ueberwachung-ist-diefalsche-antwort-a-1084629.html

[3]Frank Pasquale, The Black Box Society: The Secret Algorithms That Control Money and Information (Harvard University, 2015); Mnih Volodymyr et al., Human-level control through deep reinforcement learning, Nature 518 (2015), pp. 529–533; Kai Schlieter, Die Herrschaftsformel: Wie Künstliche Intelligenz uns berechnet, steuert und unser Leben verändert (Westend, 2015); Thomas R. Köhler, Der programmierte Mensch: Wie uns Internet und Smartphone manipulieren (Frankfurter Allgemeine Buch, 2012).

[4]Dirk Helbing et al., Digitale Demokratie statt Datendiktatur, Spektrum der Wissenschaft 1/2016 (17 December 2015), http://www.spektrum.de/news/wie-algorithmen-und-big-data-unsere-zukunft-bestimmen/1375933; see also the podcast of the event "Wie verändert die digitale Revolution unsere Demokratie?", https://www.volkswagenstiftung.de/aktuelles/aktdetnewsl/news/detail/artikel/muendige-wege-durch-dendatendschungel/marginal/4954.html; Dirk Helbing, Societal, economic, ethical and legal challenges of the digital revolution: From big data to deep learning, artificial intelligence, and manipulative technologies, in: Jusletter IT 21 May 2015.

[5]Robert Epstein/Ronald E. Robertson, The search engine manipulation effect (SEME) and its possible impact on the outcomes of elections, Proceedings of the National Academy of Sciences of the USA 112 (2015), E4512–E4521; see also Jonathan Watts/David Agren, Hacker claims he helped Enrique Peña Nieto win Mexican presidential election, The Guardian (1 April 2016), http://www.theguardian.com/world/2016/mar/31/mexico-presidentialelection-enrique-pena-nieto-hacking; Joachim Laukenmann, Wie digitale Medien Wähler manipulieren, Sonntagszeitung (10 April 2016), http://www.sonntagszeitung.ch/read/sz_10_04_2016/gesellschaft/Wie-digitale-Medien-Waehler-manipulieren-59964; Rafael Schupisser/Schweiz am Sonntag, Roboter würden SP wählen—oder: Warum Twitter eine Gefahr für die Demokratie ist, Watson (21 February 2016), http://www.watson.ch/International/Social%20Media/558554953-Roboter-würden-SP-wählen---oder--Warum-Twitter-eine-Gefahr-fürdie-Demokratie-ist

[6]Clio Andris et al., The rise of partisanship and super-cooperators in the U.S. House of Representatives, PLoS ONE 10(4): e0123507 (2015). Rising social polarisation, populism and extremism are

Despite its power, Big Nudging is not as effective as hoped in reaching its health- and environment-related goals.[7] Therefore, companies and countries like China have started to introduce Citizen Scores that rate peoples" behaviors, including the links they click in the Internet.[8] These scores will then determine the interest rates of loans offered, health benefits, travel visa, or jobs one might get. In such a way, it is easy to create a "digital nose ring" to make citizens do certain things. As a consequence, the use of Big Nudging and Citizen Scores will increasingly reduce the freedom of decision-making of citizens and with this, the ability to control their own lives. It recently becomes clear that this approach can be highly invasive: We are told to do 10,000 steps every day. The calories and kinds of food we eat become increasingly measured. The accumulated information might be used to decide who is entitled to get certain kinds of health services and who not. Since gene sequencing is cheap and gene editing is possible (e.g. with the Crispr CAS-9 method), experts, furthermore, think about eliminating diseases by "selective breeding".[9] Various countries have also built detainment camps for millions of people,[10] in order to be able to manage pandemics, future disasters, and social uprisings. The refugee camp in Idomeni is a sad example of how such camps may be operated.[11] Taking all of this together, there are currently many signs of an upcoming new totalitarism[12]—a digitally empowered political regime that would be highly privacy-invasive and potentially more dicta- torial than any political regime we have seen in the past.

One reason for this development seems to be related to a serious misunderstand- ing of what Big Data can accomplish. Since Chris Anderson's claim that "the data deluge makes the scientific method obsolete", the idea has spread that one could optimize the world and rule it like a benevolent dictator, if one just had enough

---

concerning phenomena of our time, and they can be interpreted as social cascading effects in a overly connected world (see Dirk Helbing, Globally networked risks and how to respond, Nature 497 (2013), pp. 51–59); taxing links to reduce over-connectivity and resulting systemic risks may help.

[7]Jonathan Rowson, Nudge is not enough..., Guardian (19 July 2011), http://www.theguardian.com/commentisfree/2011/jul/19/nudge-is-not-enough-behaviour-change

[8]Jay Stanley, China's nightmarish citizen scores are a warning for Americans (5 October 2015), https://www.aclu.org/blog/free-future/chinas-nightmarish-citizen-scores-are-warning-americans

[9]New CRISPR technology can be used to make designer babies (7 January 2016), http://www.news-medical.net/news/20160107/New-CRISPR-technology-can-be-used-to-make-designer-babies.aspx; see also Kemal Atlay, Gene-editing poses ethical questions, The Saturday Paper (16 April 2016), https://www.thesaturdaypaper.com.au/2016/04/16/gene-editing-poses-ethical-questions/14607288003113

[10]See https://www.youtube.com/watch?v=FfkZ1yri26s and http://info.publicintelligence.net/USArmy-InternmentResettlement.pdf

[11]See https://www.amnesty.org/en/latest/news/2016/04/greece-refugees-detained-in-dire-conditions-amid-rushto-implement-eu-turkey-deal/ and http://www.amnestyusa.org/news/press-releases/amnesty-internationalreleases-new-report-on-refugee-crisis-pushes-obama-to-do-more

[12]Frank Schirrmacher (ed.) Technologischer Totalitarismus (Suhrkamp, 2015).

data.[13] It sounds extremely plausible that "more data is more knowledge, more knowledge is more power, and more power is more success". In the meantime, however, Data Scientists have realized that there are many traps one can fall into.[14] These are not only related to the limitations of Big Data analytics (e.g. the fact that correlations do not mean causality and that more data imply more spurious patterns[15]—the well-known overfitting problem). It is also related to the fact that data volumes grow faster than processing power and systemic complexity grows even faster than data volumes.[16] Therefore, good theories are needed to decide what subset of data should be processed and how. Furthermore, there is no scientific method to determine the right goal function: should it be gross domestic product per capita (GDP) or sustainability, power or peace, happiness or life expectancy, or anything else? In modern complex societies, the common answer to this dilemma has been pluralism. We should not give up on it, because this would make our society more vulnerable: pluralism allows for diversity, which is the basis of high innovation rates, collective intelligence and societal resilience (it hedges risks and can better cope with uncertainty).[17]

---

[13]Dirk Helbing, Society is not a machine, https://www.edge.org/response-detail/26795. Note that it is often said that one needs a benevolent dictator, when decisions are time-critical and finding consensus would take too long. However, a benevolent dictator can easily make mistakes, particularly in a complex world. If the benevolent dictator is powerful, such mistakes will be big and affect the existence of millions of people. Therefore, I suggest that different solution approaches should be tried out in various places, as the federal and subsidiarity approaches suggests, and that these experiments should be scientifically evaluated to spread and (co-)evolve the best solutions. In a complex world, such a pluralistic approach is more promising for humanity to master difficult times than applying big solutions to all. By the way, such a pluralistic approach has just been used by the Daimler AG to identify good strategies for the future, see https://www.youtube.com/watch?v=cpuz8x-6BDA

[14]David Lazer et al., The Parable of Google Flu: Traps in Big Data Analysis, Science 343 (2014), 1203–1205; Dirk Helbing, Thinking Ahead—Essays on Big Data, Digital Revolution and Participatory Market Society (Springer, Berlin, 2015).

[15]For an entertaining illustration of the problem see the book by military intelligence analyst Tyler Vigen, Spurious Correlations (Hachette, 2015).

[16]Dirk Helbing, What the digital revolution means for us, Science Business (12 June 2014), http://www.sciencebusiness.net/news/76591/What-the-digital-revolution-means-for-us

[17]Dirk Helbing/Evangelos Pournaras, Build Digital Democracy, Nature 527 (2015), 33–34, http://www.nature.com/news/society-build-digital-democracy-1.18690

I am writing all this, because democracies worldwide have repeatedly been questioned[18] and come under pressure.[19] In Europe, we have seen this in Hungary, Poland, France, and Turkey, but also elsewhere.[20] In fact, it is now possible to build data-driven versions of well known historical political systems: fascism 2.0, communism 2.0, feudalism 2.0, capitalism 2.0, and democracy 2.0. Our societies are at a crossroads (or, scientifically speaking, at a tipping point). As a consequence, we should make up our minds and take a conscious decision. We need a public debate to determine the path we want to take. In the following, I will try to sketch some implications of the various data-driven models mentioned above.

Fascism 2.0—the Big Brother and Brave New World Society: This system is turning majority opinions into social norms, laws, and regulations with little protection of human rights, particularly those of minorities. It leads to a populist governance supported by a propaganda apparatus that is controlled by government and business elites. In the beginning, resources of minorities are redistributed to the majority, which increases the popularity of the government. Eventually, however, the loss of diversity leads to a lack of innovation, which undermines the success of the economy and the functionality of the society. To counteract the resulting destabilization of the socio-economic system and overcome the shortage of resources, fascist societies typically end up in wars. They also tend to value diverse people differently (e.g. to attribute a lower value to certain ethnicities, groups or religions), which can lead to ethnic cleansing and social sorting. A trend towards fascism 2.0 can be seen in several countries today, also in Europe.

Communism 2.0—some are calling it the Big Mother Society: This system is pursuing a benevolent dictator approach, trying to optimize the state of society. The "caring state",[21] which engages in Big Nudging, clearly has elements of this.

---

[18]Tony Blair, Is Democracy Dead? The New York Times (4 December 2014), http://www.nytimes.com/2014/12/04/opinion/tony-blair-is-democracy-dead.html; Harald Welzer, Die Demokratie—ein Auslaufmodell, Die Welt (2 August 2008), http://www.welt.de/welt_print/article2332799/Die-Demokratie-ein-Auslaufmodell.html; Jakob Tanner, Demokratie—ein Auslaufmodell?, Tagesanzeiger (19 July 2015), http://www.tagesanzeiger.ch/schweiz/standard/Demokratie-ein-Auslaufmodell/story/20251334; Michael Safi, Have millennials given up on democracy?, The Guardian (18 March 2016), http://www.theguardian.com/world/2016/mar/18/have-millennials-givenup-on-democracy

[19]In his remarks by at the White House Correspondents' Dinner on 30 April 2016, President Obama said: "[...] this is [...] a time around the world when some of the fundamental ideals of liberal democracies are under attack, and when notions of objectivity, and of a free press, and of facts, and of evidence are trying to be undermined. Or, in some cases, ignored entirely. And in such a climate, it's not enough just to give people a megaphone. [...] that's why your power and your responsibility to dig and to question and to counter distortions and untruths is more important than ever." See https://www.whitehouse.gov/the-press-office/2016/05/01/remarks-president-whitehouse-correspondents-dinner. Also see Carsten Könneker, Fukushima der Künstlichen Intelligenz, http://www.spektrum.de/news/interview-die-unterschaetzten-risiken-der-kuenstlichen-intelligenz/1377620.

[20]Harald Welzer, Die smarte Diktatur. Der Angriff auf unsere Freiheit (Fischer, 2016).

[21]Richard H. Thaler/Cass R. Sunstein, Nudge: Improving Decisions about Health, Wealth, and Happiness (Penguin, 2009).

Communism 2.0 imposes values, norms, and forms of life on people. Communism 2.0 engages in a centralized, top-down planning of the use of resources. In this process, the goals are set by the government. In many cases, this includes the re-distribution of resources from certain elites to a broader public. However, communism 2.0 undermines competition, innovation and entrepreneurship, thereby reducing the amount of resources available, ending up in a desolate economic situation that requires the government to ration resources.

Feudalism 2.0—the Big Other Society (called surveillance capitalism by some people)[22]: This system amasses huge amounts of customer data and basically turns citizens into products. The system is based on the accumulation of resources and power in the hands of a small business elite, which is said to be in favor of an efficient use of resources ("economies of scale").However, there are also undesirable side effects such as the misuse of power, relatively low innovation rates, and too-big-to-fail problems (as we have seen them in the financial sector).

In some economic sectors, the accumulation of resources and power has created quasimonopolies. Typically, the related monopolists, oligarchs or plutocrats demand the right of breaking the rules[23] (which is often framed as "creative destruction") and to determine the rules of the future.[24] Some of Silicon Valley's IT giants openly admit that they consider democracy an "outdated technology".[25] It appears that they

---

[22]Shoshana Zuboff, Big Other: Surveillance Capitalism and the Prospects of an Information Civilization, Journal of Information Technology 30 (2015), 75–89. It is a well-known problem that even bottom-up organizational approaches tend to turn into feudalistic structures eventually (see the Iron Law of Oligarchy, https://en.wikipedia.org/wiki/Iron_law_of_oligarchy). This has also be found for Wikipedia (see http://www.sciencealert.com/wikipedia-is-basically-just-another-old-fashioned-bureaucracy-study-finds). To counteract this tendency, various measures can be taken, including: decentralization, separation of roles, limited office periods, democratic elections, random elements, multi-dimensionality/diversity/pluralism, mechanisms avoiding too much accumulation of resources and power, division of power/checks and balances, participatory approaches, shared values, accountability, and transparency. Such elements should be built into the system ("democracy by design").

[23]i.e. applicable laws or even the constitution.

[24]The diversity of solutions. Requiring transparency, openness, and the removal of obstacles to interoperability and innovation seems to make more sense. Another issue is the following: to expand quickly, big businesses focus on products and services generating the highest revenues (say, 20%). Other products and services, even if profitable, are often not provided. SMEs, but also non-government organizations and citizen engagement are, therefore, needed, to create and offer products and services, which are desirable, but less profitable or require even an investment. Such products and services are important for a thriving society as well. In other words, big business can efficiently satisfy the basic needs of millions or billions, but this alone will typically not create a high quality of life. For this, a close-knit information, innovation, production and service ecosystem is needed, which is highly differentiated and diverse. SMEs are indispensable.

[25]The complete quote is: "Die Demokratie ist eine veraltete Technologie. [. . .] Sie hat Reichtum, Gesundheit und Glück für Milliarden Menschen auf der ganzen Welt gebracht. Aber jetzt wollen wir etwas Neues ausprobieren." Another quote from Larry Page is: "Es gibt eine Menge Dinge, die wir gern machen würden, aber leider nicht tun können, weil sie illegal sind. Weil es Gesetze gibt, die sie verbieten. Wir sollten ein paar Orte haben, wo wir sicher sind. Wo wir neue Dinge ausprobieren und herausfinden können, welche Auswirkungen sie auf die Gesellschaft haben."

want to replace it by a new, data-driven, worldspanning operating system[26] to overcome the limitations of nation-based politics.[27]

All three of the above governance systems (i.e. fascism 2.0, communism 2.0, and feudalism 2.0) are run with Big Data and Artificial Intelligent technologies (smart algorithms). They are all operated in a top-down way and imply severe breaches of basic democratic principles and achievements that many societies have fought for over thousands of years. I will now continue with a discussion of novel, data-driven forms of society that have strong bottom-up components: democracy 2.0 and capitalism 2.0. Here, "big solutions" are replaced by the "glocality" principle: "think global, act local".

Democracy 2.0: This system recognizes the important role of pluralism and diversity for high innovation rates, societal resilience, and collective intelligence. It understands that socioeconomic and cultural diversity is just as important as biodiversity, which requires a proper protection of minorities and human rights (including informational self-determination). It is committed to a division/separation of powers as a means of balancing the interests of different stakeholders and ensuring peace. It also offers participatory opportunities for citizens in recognition of the importance of civic engagement for a thriving society. It provides freedoms to citizens in exchange for responsible behavior, and therefore invests in the education and enlightenment of its citizens.[28]

See https://www.youtube.com/watch?v=NCtBdVpUY08 and Christoph Keese, Silicon Valley: Was aus dem mächtigsten Tal der Welt auf uns zu kommt (Albrecht Knaus, 2014). One such place beyond the rule of national law is CERN. It has built something like a Crystal Ball for society, which uses predictive analytics, but also predictive programming technologies ("big nuding"), and machine learning approaches (Artificial Intelligence). See Peter Welchering, Die Software vom Cern spielt Orakel, FAZ (25 April 2016), http://www.faz.net/aktuell/technik-motor/computer-inter net/einsatz-der-cern-prognose-software-im-alltaeglichen-bereich-14184322.html; Chris Merriman, CERN: The gulf between machine learning and AI, The INQUIRER (29 July 2015), http://www. theinquirer.net/inquirer/feature/2419669/cern-the-gulf-between-machine-learning-and- artificialintelligence. It is now important to create a scientific, ethical and governance framework, which is pluralistic, interdisciplinary, transparent, and accountable, to prevent improper use of these technologies.

[26]Adrian Lobe, Google will den Staat neu programmieren, FAZ (10 October 2015), http://www.faz. net/aktuell/feuilleton/medien/google-gruendet-in-den-usa-government-innovaton-lab-13852715. html; Katherine Noyes, Forget Trump and Clinton: IBM's Watson is running for president, PC World (8 February 2016), http://www.pcworld.com/article/3031137/forget-trump-and-clinton- ibms-watson-is-running-forpresident.html

[27]Thomas Schulz, Die Weltregierung: Wie das Silicon Valley unsere Zukunft steuert, Spiegel (4 March 2015), http://www.spiegel.de/international/germany/spiegel-cover-story-how-silicon-val ley-shapes-our-future-a-1021557.html

[28]Theda Skocpol/Morris P. Fiorina (eds.) Civic Engagement in American Democracy (Brookings, 1999); Aaron Smith, Civic engagement in the digital age, PewResearchCenter (25 April 2013) http://www.pewinternet.org/2013/04/25/civic-engagement-in-the-digital-age/; Essop Pahad, Politi- cal Participation and Civic Engagement, Progressive Politics Vol 4.2 (1 July 2005), http://www. policy-network.net/uploadedfiles/publications/publications/pahad-final.pdf

The above principles may be seen as the lessons learned over centuries—from many wars and revolutions. As the complexity and diversity of socio-economic systems has increased over historical periods of time, overall, their degree of participation has increased as well. Democracy may, therefore, be seen as a governance form enabling diversity and turning it into public and private benefits.[29]

Of course, high levels of complexity and diversity also imply significant challenges. However, these can now be addressed by personal digital assistants helping individuals and companies to coordinate their activities and interact more successfully.[30] Such digital assistants may use Artificial Intelligence technology, but act on behalf of their users rather than trying to control them on behalf of companies or governments.[31] The important point is to build digital platforms, which support efficient cooperation, online deliberation, and collective intelligence. Supporting better decision-making, responsible (inter)action, self-organization, and self-regulation will be key to eGovernance. I will discuss this in more detail below.

Capitalism 2.0: This system is built on liberalism, i.e. civil and entrepreneurial freedom in favor of high innovation rates. It is largely organized in a bottom-up way, i.e. based on selforganization and self-regulation. It is, therefore, flexible, efficient, adaptive and resilient. Strong reward mechanisms for inventions and creative products or services are the fuel of the success of this system.

A novel aspect of capitalism 2.0 is the overwhelming importance of network interactions. As a consequence, everything we do has side effects, feedback effects, or cascading effects, and "systemic thinking" or—differently phrased—"ecosystems thinking" becomes key to longterm (sustainable) success.[32] In fact, the nations with the most diverse product space are those that thrive most.[33] "Ecosystems thinking" also implies new forms of competition, which are combined with cooperation (so-called "co-opetition"). This is reflected by the important roles that cocreation, co-evolution, and collective intelligence play in the now emerging digital economy.[34] As I will elaborate later, this new form of capitalism will be catalyzed by a

---

[29]Timothy D. Sisk (ed.) Democracy at the Local Level, Chap. 3 (2010) http://www.idea.int/publications/dll/; Nico Stehr, Climate policy: Democracy is not an inconvenience, Nature 525 (24 September 2015), 449–450.

[30]Dirk Helbing, Interaction Support Processor (2015), https://patentscope.wipo.int/search/en/detail.jsf?docId=WO2015118455

[31]Elon Musk: "I think the best defense against the misuse of AI is to empower as many people as possible to have AI. If everyone has AI powers, then there's not any one person or a small set of individuals who can have AI superpower." See http://www.kurzweilai.net/musk-others-commit-1-billion-to-non-profit-ai-research-company-tobenefit-humanity

[32]As a side effect, the concept of the "homo economicus" will increasingly be replaced by the concept of "homo socialis". See the Special Issue on Homo Socialis, Review of Behavioral Economics, Vol. 2 (2015), http://www.nowpublishers.com/article/Details/RBE-0032

[33]Cesar A. Hidalgo et al., The Product Space Conditions the Development of Nations, Science 317 (2007), 482–487.

[34]Dirk Helbing, The Automation of Society Is Next: How to Survive the Digital Revolution (CreateSpace, 2015).

new kind of financial system, which will fix current environmental, social, and unsustainability issues: "socio-ecological finance" (or "finance 4.0").

11.1 The Perfect Storm: Resource Shortages and Other Crises

Before I discuss what governance approach we should choose, we first need to discuss the situation our society is faced with. Obviously, we are confronted with an unsolved financial, economic and spending crisis, with unstable peace, with climate change, and with cyber threats, to mention just a few of our existential issues. A closer analysis reveals that they may all be rooted in a single problem: our unsustainable way of life. Industrialized countries are currently consuming 3.5–4.5 times the amount of renewable material resources. For some time, they managed to keep up their lifestyle by means of globalization, which made resources from other countries accessible. Now, however, globalization has reached its limits, as the exploitation of other celestial bodies (such as asteroids or Mars) is not yet possible.

Already in 1972, when the Club of Rome published "The Limits to Growth",[35] it became clear that we might run into trouble. The analysis triggered controversial debates, but it was followed up and basically confirmed by the Global 2000 report issued by the US president, and it was further refined by an update.[36] Initially, these studies have fueled an environmental movement towards the protection of our planet and more responsible consumption. Then, however, industry and public media kept suggesting for decades that there was no problem, supporting excessive consumerism. If we were running short of a certain resource, it was argued, engineers would come up with a solution that would fix the problem. The same circles kept pointing out that there would be no man-made climate change problem. This allowed them to do business as usual.

The "Limits to Growth" and Global 2000 studies particularly stressed that there was an impending oil shortage and an overpopulation problem. By the end of the twenty-first century, our planet would host significantly less people (more than one billion people less than during the period of highest population, which is expected to occur around the year 2030). Back in 1973, during the first oil crisis, many public roads were closed during weekends (for example, in Germany). However, since then it appears to the naive consumer that the problems have been solved. In many areas of the world, energy consumption has even doubled. In fact, we are not yet running out of oil, but problems are expected to occur as soon as the global production rate of

---

[35]Donella H. Meadows, Limits to Growth (Signet, 1972); Donella H. Meadows et al., Limits to Growth: The 30- Year Update (Chelsea Green, 2004).

[36]Gerald O. Barney, The Global 2000 Report to the President, http://www.geraldbarney.com/G2000Page.html; Gerald O. Barney, Global 2000 Revisited, http://www.geraldbarney.com/G2000Revisit.html

oil goes down or oil production gets less efficient, i.e. more energy and effort is required to get it out of the soil.[37] The implications become clear when considering that most of our economy today depends on oil. It fuels production processes and transportation worldwide. It is used to produce plastic and other materials. It is required to produce fertilizer, i.e. it is needed for global food production. In fact, the world population has increased by a factor of 6 in the past 150 years as a result of the oil-based economy. So, if we would run short of oil or have to use less of it (e.g. to mitigate climate change), a large number of people might die,[38] which explains why nobody wants to talk about this. And oil is not the only problem.[39]

There are many signs that the elites knew about the issues and tried to solve them, while the wider public and even many researchers were unaware of their true dimensions.[40] Entirely new billion-dollar industries were built in attempts to overcome the existing and anticipated issues. Unfortunately, none of them really fixed the world's problems—each of them created new ones. The automobile industry managed to significantly reduce fuel consumption per 100 kilometers, but the number of cars has dramatically increased. A similar thing applies to air traffic. Each flight saves a lot of fuel, but the number of passengers and miles travelled increased a lot. Electricity production turned from oil-based to nuclear technologies, but besides the risk of nuclear accidents (as in Chernobyl and Fukushima), two problems have not been solved: the longterm storage of nuclear waste and the threat of nuclear terror attacks ("dirty bombs"). Nuclear fusion (as it is, for example, attempted by the ITER reactor) would potentially solve these issues, but it is still in an early stage of development—far from industrial-scale energy production. Long pipelines have been built to replace oil by gas. However, deliveries in Europe suffer from political uncertainties (the conflict with Russia and the political instability in North Africa). Besides, burning gas contributes to climate change, which endangers one sixth of all species[41]—that would be the biggest mass extinction since the death of dinosaurs. Worse comes to worse, it endangers the global food system, too.

Solar power is the great hope, but depends on regions with a lot of sunshine hours such as Africa. In fact, the DESERTEC project wanted to invest 400 billion Euros to produce energy for entire Europe, but the project collapsed due to the political turbulences caused by the Arab Spring. The Arab Spring was triggered by high food prices, which were a side effect of biofuel production (which started to compete

---

[37]This is sometimes phrased as "peak oil" problem.

[38]Roy Scranton, Learning to Die in the Anthropocene (City Lights, 2015).

[39]Bundesamt für Umwelt BAFU, Umweltzustand: Globale Megatrends, http://www.bafu.admin.ch/umwelt/12492/12803/index.html?lang=de, see also http://visual.ly/born-2010-how-much-left-me

[40]I am convinced that public awareness and involvement must be largely increased, if we want to be able to come up with better solutions.

[41]Sid Perkins, Climate change could eventually claim a sixth of the world's species, Science (30 April 2015), http://www.sciencemag.org/news/2015/04/climate-change-could-eventually-claim-sixth-world-s-species

with food production), combined with financial speculation.[42] Unfortunately, bio-fuel production also contributes to deforestation and land erosion. Using special bacteria to turn unused biomass into fuel appears promising, but does not seem to reach sufficient production rates so far. Furthermore, large wind parks have been built to generate electricity, but electricity production is spiky, i.e. it requires efficient storage, which has been lacking in many wind-rich regions. Novel battery systems to store electrical energy seem to provide a solution now. However, it is not quite clear how efficient and environmentally friendly the production and recycling of such battery systems is.[43] The recent use of fracking technology looks like a desperate attempt to prolong the era of the oil-based economy: it has serious side effects on the environment and health,[44] and causes man-made earthquakes. Despite all this, I am confident that energy production will not be the main resource bottleneck in future. We need to understand, however, that there is not one solution that fixes all our energy problems, if we would just roll it out globally. Instead, we need to engage in diverse approaches, in local energy generation (particularly of solar and geothermal energy), in matching energy production and consumption better (with so-called "smart grid" solutions), and in a new lifestyle (as discussed below).

Of course, energy is not the only thing that counts. In other economic sectors, we have made much less progress with the buildup of a low-carbon economy. For example, the food, chemical and pharmaceutical industries largely depend on oil. In particular, oil is needed for the production of fertilizer. Fertilizer also depends on nitrogen and phosphorus—two elements human are currently overusing. Further-more, soil is degrading in many areas of the world and becoming more arid. Water shortages are impending. This is partly due to intensive meat production, which implies further problems (not just ethical ones): it makes antibiotics less effective. Multi-resistant bacteria are spreading and may cause a global pandemic of pre-antibiotic scale, killing millions of people. The amount of vegetables produced has also increased a lot. Part of this success is based on genetically modified organisms (GMO). However, some of the most used poisons in agriculture seem to have unwanted side effects. Insect species, in particular bees that are enormously important for food production, have been dramatically reduced.[45] In humans, the rates of obesity, diabetes and various kinds of cancer have significantly increased. It

[42]Damian Carrington, Are food prices reaching a violent tipping point, The Guardian (25 August 2011), http://www.theguardian.com/environment/damian-carrington-blog/2011/aug/25/food-price-arab-middleeast-protests; Yaneer Bar-Yam/Greg Lindsay, The real reason for spikes in food prices, Reuters (25 October 2012), http://blogs.reuters.com/great-debate/2012/10/25/the-real-reason-for-spikes-in-food-prices/

[43]Danny King, Tesla Model S fined for excessive emissions in Singapore, AutoBlog (8 March 2016), http://www.autoblog.com/2016/03/08/tesla-model-s-fined-for-excessive-emissions-in-singapore/

[44]See, for example, http://www.focus.de/gesundheit/news/gefaehrliche-krankheit-ein-dorf-lebt-in-angst-krebs-istdie-einzige-todesursache_id_5443253.html

[45]Dramatisches Insektensterben, NABU (13 January 2016), https://www.nabu.de/news/2016/01/20033.html

is certainly true that humans managed to reduce some of the world's hunger and to increase life expectancy, but it is far from obvious how we can cover the increasing costs of our health system and sustain the world's population, which has quickly grown.

The financial industry is in trouble, too. Some people are convinced that the financial crisis in 2007/08 was triggered by skyrocketing oil prices, which made the production and consumption of certain products increasingly unaffordable.[46] Others see demographic reasons.[47]

The use of anti-baby pills, they say, has reduced the size of the younger population, so that there are less people who can invest in stocks and real estate. In fact, the financial meltdown started with an overvalued housing market in California.[48] This eventually triggered a financial crisis around the globe, an economic crisis, and a public spending crisis, which we have not managed to overcome yet. As a result, inequality has largely increased. It will actually soon reach levels greater than before the French revolution, which raises concerns about the possibility of social instability and political unrests.[49]

Some may argue that this inequality reduces the pressure on scarce resources. However, today's kind of capitalism cannot work without perpetual growth (which is a side effect of having to pay interest rates for loans). Consequently, the current economic system is in a danger to collapse. The Federal Reserve, the European Central Bank and other National Banks are literally pumping trillions of dollars into the economy to prevent this from happening. However, the money did not boost new production sites, companies and business models to the extent needed in order to create a booming economy. Instead, there are large unemployment rates in many countries, particularly among young people. The OECD, the WEF and the IMF are all pointing out that inequality has reached a level that harms our economy.[50]

Currently, 62 people own as much as the poorest half of the world population.[51] I do not have objections against rich people or inequality, in principle, but when small- and medium-sized companies disappear, the backbone of society is

[46]Justin Lahart, Did the oil price boom in 2008 cause crisis? The Wall Street Journal (3 April 2009), http://blogs.wsj.com/economics/2009/04/03/did-the-oil-price-boom-of-2008-cause-crisis/

[47]Demographics: What variable best predicts a financial crisis? (16 July 2010), http://andrewgelman.com/2010/07/16/demographics_wh/

[48]Financial crisis of 2007–08, https://en.wikipedia.org/wiki/Financial_crisis_of_2007-08

[49]Nick Hanauer: Beware, fellow plutocrats, the pitchforks are coming (12 August 2014), https://www.youtube.com/watch?v=q2gO4DKVpa8

[50]Inequality hurts economic growth, finds OECD research (12 September 2014), http://www.oecd.org/newsroom/inequality-hurts-economic-growth.htm; Carter C. Price, Why inequality harms economic growth (11 December 2014), https://www.weforum.org/agenda/2014/12/why-inequality-harms-economic-growth/; Phillip Inman, IMF study finds inequality is damaging to economic growth, The Guardian (26 February 2014), http://www.theguardian.com/business/2014/feb/26/imf-inequality-economic-growth

[51]Larry Elliott, Richest 62 people as wealthy as half of world's population, says Oxfam, The Guardian (18 January 2016), http://www.theguardian.com/business/2016/jan/18/richest-62-billionaires-wealthy-half-worldpopulation-combined

damaged.[52] The current problem is that the average spending power has become so low that many firms cannot sell their products well anymore. This is bad for companies and the economy. It is bad for people and our society. To make us spend more money, interest rates have been steadily reduced by the FED and central banks around the world, but instead of consuming more, people are losing their savings for difficult times,[53] which fuels populist movements.[54] In the meantime, the interest rates have effectively reached a value of zero, more or less. This means that capitalism 1.0, which was built on competitive mechanisms (such as interest rates) and on mass consumption, is basically dead.[55]

In other words: our current system is broken, it will not work much longer. Attempts to cement the outdated world order of the twentieth century with secret agreements may be highly counterproductive for the future of our world.[56] To keep people in check and society under control, governments have built powerful surveillance and control system. We may also see a rationing of resources. Such a system would correspond to Communism 2.0 (if organized by governments) or Feudalism 2.0 (if organized by multi-national corporations). In more serious

---

[52]Stéphanie Thompson, The digital revolution could destroy the middle class, warns Joe Biden, World Economic Forum (21 January 2016), https://www.weforum.org/agenda/2016/01/the-digital-revolution-could-destroy-themiddle-class-warns-joe-biden/; "Middle-class Joe" Biden tells Davos bosses to look after workers, Fortune (20 January 2016), http://fortune.com/2016/01/20/joe-biden-davos-workers/

[53]For Germany, this effect sums up to 125 billion EUR, see http://www.focus.de/finanzen/news/ezb-hat-denbogen-ueberspannt-wegen-niedrigzins-deutsche-sparer-haben-100-milliarden-euro-in-fuenf-jahren-verloren_id_5442246.html; other numbers speak even of 327 billion EUR, see http://m.welt.de/finanzen/geldanlage/article153466283/Niedrigzins-kostet-Deutsche-327-Milliarden-Euro.html

[54]Thomas Fricke, Aufstieg der Rechtspopulisten: Schaut auf die Banken, Spiegel (15 April 2016), http://www.spiegel.de/wirtschaft/aufstieg-der-rechtspopulisten-liegt-an-der-finanzkrise-kolumne-a-1087139.html

[55]Paul Mason, The end of capitalsm has begun, The Guardian (17 July 2015), http://www.theguardian.com/books/2015/jul/17/postcapitalism-end-of-capitalism-begun; Carolin Haentjes, Der Kapitalismus ist am Ende, Der Tagesspiegel (6 April 2016), http://www.tagesspiegel.de/kultur/paul-mason-im-hkw-berlin-der-kapitalismusist-am-ende/13412300.html

[56]Of course, international trade and service agreements can make a lot of sense, but they may also amplify existing problems and inequality, too: if the roles of customers and citizens are not strengthened, such agreements will reduce the likelihood of new solutions and accelerate the accumulation of formerly public or widely spread property and power in the hands of a few big companies, promoting Feudalism 2.0. This seems to be one of the great public concerns against the CETA, TTIP and TISA agreements, besides the possible weakening of democracy as well as social, environmental, and legal standards. Note that replacing the precautionary principle for new products by a risk management approach, as it is demanded by the USA (see http://www.greenpeace.org/euunit/en/News/2016/TTIPleaks-confidential-TTIP-papers-unveil-US-position/), has at least two drawbacks: (1) It often takes decades from the first evidence of a serious product risk until this finding has been established as a widely recognized fact; during this long time period, much harm can occur. (2) If risk management is combined with big and often global solutions, a mistake can easily become a big mistake, from which it may be hard to recover (such as the overuse of carbon-based energy resources).

scenarios, a nine- or ten-digit number of people could die from hunger, disease or war. The scenarios discussed in connection with future resource shortages are depressing and sad. It is increasingly obvious that our economy will run in an evolutionary dead end, if we go on as before. We can certainly not allow this to happen, given that the lives of so many people are at stake. Before I explain, how we can mitigate our problems by means of democracy 2.0 and capitalism 2.0, I will discuss, how we got ourselves into so much trouble that we must probably talk of a "systemic failure".

11.2 Why Our Socio-economic System Fails

Multiple institutional failures are actually quite typical in times when a society undergoes a transition from one historical age to another. We have seen this in the past, when the agricultural society was replaced by the industrial society, and when the industrial society became the service society. Now, we are in the middle of a transition to the digital society. History tells us that these transformations tend to come along with financial and economic crises, with revolutions and wars, but after the respective transformation has been accomplished, further growth occurs. This time, we need to be smarter. We must avoid another revolution, war, or holocaust. We must reinvent and re-organize our socio-economic system before it collapses.

While the agricultural society was ruled in a top-down way by kings, the industrial society was self-organized in a bottom-up way by entrepreneurs, and the service society was run by administrations. I am convinced that the digital society will be based on collective intelligence. Before this can happen, however, we need to create a suitable framework for it (democracy 2.0 and capitalism 2.0, as I will argue). If we go on as before, however, our problems will further intensify, until systemic collapse is inevitable.

It is important to understand that this collapse can result even when everyone applies the established success principles of the past and has the very best intentions. Politicians, for example, may try their best to listen to the interests of people, but they can talk to just a few of them. Therefore, they will most likely talk to people representing a lot of other people, for example, industrials. This will automatically lead to politics that is oriented primarily at economic interests. Particularly in times of scarcity, this seems to be the right thing to do. If Maslov's pyramid of needs were right, food and shelter would be the main thing to care about, and everything else would become secondary or even less important.[57]

Consequently, the job of entrepreneurs would be to minimize the use of resources, which seems to speak for committing to "economies of scale", i.e. mass production and monopolies. In fact, every industrial sector has learned to decrease

---

[57]I would like to call this pyramid of needs into question: information, friendship and solidarity experience all-time heights in harsh times, too.

the use of valuable resources per unit delivered. Cars and planes run on less fuel. A lot of supermarket food contains less and less ingredients that we would have considered food 40 years ago.[58] Similar trends can be observed in other economic sectors.

Today, more than a billion people have more comfortable lives than kings used to have just a few generations ago. Thanks to the access to global resources and rationalization, the economy grew and grew. The public media helped to produce the culture of materialism and consumerism, which made it happen. In fact, our economy had to grow, because it builds on loan-based investments that require lenders to pay an interest rate to the banks. Banks had to take an interest rate to make money and exist. Scientists and engineers invented as told within the framework of the current system. They largely worked on subjects, which allowed them to raise third-party funding: subjects that companies cared about.

All in all, everyone did what they were supposed to do,[59] but the economy, politics, science, and our lives were more and more dominated by one single dimension: economic efficiency.[60] Nevertheless, since the financial crisis, this economic system has come to its limits.

Globalization does not progress anymore.[61] Central banks pump trillions of dollars into the markets, but this money does not reach the average customer and does not boost real investments as expected. So far, all desperate attempts to restart the engine of the world economy have failed.

The game seems to be over. After the next financial and economic crisis, which some people expect to come pretty soon, we will probably face similar conditions as in the 1930s.

Politics has already prepared for this: by investing in surveillance systems, weapons and detainment camps, but also in new kinds of propaganda and censorship

---

[58]Hannes Grassegger, China auf der Zunge, ZEIT ONLINE (15 November 2013), http://www.zeit.de/wirtschaft/2013-11/china-auf-der-zunge-essen-kochen;   Gesundheitsrisiko: Europol   findet Rekordmenge gefälschter Lebensmittel, Spiegel (30 March 2016), http://www.spiegel.de/wirtschaft/service/europol-findet-rekordmengegefaelschter-lebensmittel-a-1084739.html

[59]In a letter dated 22 July 2009 to the Queen of England, the British Academy also came to the conclusion that evenwell-intended behavior may lead to systemic failure: "When Your Majesty visited the London School of Economics last November, you quite rightly asked: why had nobody noticed that the credit crunch was on its way? [. . .] So where was the problem? Everyone seemed to be doing their own job properly on its own merit. And according to standard measures of success, they were often doing it well. The failure was to see how collectively this added up to a series of interconnected imbalances over which no single authority had jurisdiction. [. . .] Individual risks may rightly have been viewed as small, but the risk to the system as a whole was vast. [. . .] So in summary [. . .] the failure to foresee the timing, extent and severity of the crisis [. . .] was principally the failure of the collective imagination of many bright people to understand the risks to the systems as a whole." See http://wwwf.imperial.ac.uk/{}bin06/M3A22/queen-lse.pdf

[60]The minimization of the use of costly resources ("rationalization") is driven by the desire to maximize revenues. However, the impact on human, social, and environmental resources, particularly public ones, is often neglected.

[61]The exploitation of celestial bodies is planned, but is not possible yet.

systems to keep people in check—every one of us (using big nudging, social media filters, and social bots,[62] for example).

The Sustainable Development Agenda of the United Nations does not sufficiently distance itself from such measures. In fact, the Agenda 2030 calls for "strong institutions".[63] One should keep in mind, however, that strong institutions can be very harmful, if power falls into wrong hands or inappropriate actions are taken.[64] For example, in a state of emergency, it often happens that measures not legitimated by the public, by science or previous success, are applied, but it is impossible to prevent them.

A politics of empowerment, in contrast, which enables citizens to help themselves, to help each other, and to contribute to solutions (by means of democracy 2.0 and capitalism 2.0, for example) seems to be more promising to master the challenges of the future in a reasonable way. In particular, the past has shown that narrowing down the solution approach to one criterion (or a few) such as security or economic efficiency, deteriorates the situation and causes a socioeconomic system to fail.

11.3   Future Scenarios

The question is, what will happen now? Several scenarios spring to mind:

Global war (WW 3): One possible and not entirely unlikely scenario is global war. Besides atomic war, we may see biological warfare (through the spread of deadly diseases), chemical warfare, or digital warfare (cyber war to make the information infrastructure dysfunctional). Each of these variants would be similarly devastating. Digitalwarfare seems to be less destructive, but would actually cause societal collapse through cascading failures.[65] As a result of any of these scenarios, hundreds of millions of people could die—it would be a true Armageddon.

---

[62]Joachim Laukenmann, Wie digitale Daten Wähler manipulieren können, Die Welt (20 April 2016),   http://www.welt.de/wissenschaft/article154572957/Wie-digitale-Daten-Waehler-manipulieren-koennen.html; see also Joachim Laukenmann, Wie digitale Medien Wähler manipulieren, Sonntagszeitung (10 April 2016), http://www.sonntagszeitung.ch/read/sz_10_04_ 2016/gesellschaft/Wie-digitale-Medien-Waehler-manipulieren-59964;   Christian   Meier/Jennifer Wilton, Social bots: Maschinen übernehmen die Macht im Internet, Die Welt (11 April 2016), http://www.welt.de/wirtschaft/webwelt/article154223388/Maschinen-uebernehmen-die-Machtim-Internet.html

[63]UN Sustainable Development Goals, http://www.un.org/sustainabledevelopment/peace-justice/

[64]Warnings of the totalitarian potential also come from elected parliamentarians in various countries, see e.g. https://www.youtube.com/watch?v=sES6_OXPwOU; further information about the Agenda 21 and Agenda 2030 can be easily found in various YouTube channels.

[65]Marc Elsberg's book "Blackout" illustrates how such cascading scenarios might unfold. For scientific analyses of cascade effects see Dirk Helbing/Hendrik Ammoser/Christian Kühnert, Disasters as extreme events and the importance of network interactions for disaster response management, in: Extreme Events in Nature and Society (Springer, 2010), pp. 319–348.

Global pandemic: A deadly disease may naturally emerge and spread. In fact, epidemiologists are expecting a large-scale pandemic already for some time. A new concern is the spread of multi-resistant bacteria. In other words, antibiotics become increasingly ineffective, as they are overused in industrial-scale meat production, which allows bacteria to adapt.

Euthanasia: Based on some criteria, it would be decided whose life will be terminated (or not prolonged, e.g. by refusing medical aid). Some experts consider having computers or "superintelligent systems" make and execute such decisions, for example, with the help of implants.

Revolution: Economic and political instability may lead to social unrest and revolutions. Elites, which have failed to live up to their promises, and their immediate beneficiaries would be killed or imprisoned for corruption. Their accumulated resources would be redistributed to relieve the scarcity of resources. If the resulting political system would use material resources more sustainably, a lot more people may survive and live on.

Citizen score: When running short of resources, a supercomputer using artificial intelligence technology would be employed to decide who would get how much of what resource, given previous behaviors, merits, and usefulness for society. Such citizen scores are already used in China to determine the payment conditions of loans, the kinds of jobs one may get, and the travel visa to other countries. The citizen score depends not only on consumption patterns, but also on the political attitude (as documented by the Internet links clicked) and by the behavior of family and friends.[66] In effect, such a system would create a caste society: some would get everything they want, others would have no chance to get certain kinds of resources.

Basic income[67]: In order to counter the rise of populism, extremism and social unrest, governments may decide to pay everyone a basic income, and allow them to earn an additional income by paid work. This would substantially reduce peoples" fear of losing their jobs (which may happen due to the quick spread of intelligent machines and automation). It would also allow people to experiment with new kinds of production and business models. Many people care more about a meaningful life than about a high material standard of living. A lot of them would turn to creative and innovative work. The demand for material resources would go down considerably, if we had a new Zeitgeist focused on the immaterial sides of life. After all "the best things in life are free."

I believe that basic income (plus a new Zeitgeist) is the best and only acceptable (responsible, legitimate) solution among the scenarios discussed above. It comes

---

[66] A similar system, called Karma Police, exists in Great Britain, see Ryan Gallagher, Profiled: From radio to porn, British spies track Web user's online identities, The Intercept (25 September 2015), https://theintercept.com/2015/09/25/gchq-radio-porn-spies-track-web-users-online-identities/

[67] Depending on the implementation, the resulting system might be considered as socialism 2.0 or social market economy 2.0.

closest to "business as usual", but this is also the greatest drawback. Below I will, therefore, propose a considerably better, data-driven solution. However, as this solution is not yet ready for use though and still needs to be implemented, we may need an intermediate solution to bridge between today's system and the future system. "Helicopter money", as it is currently discussed by economic and banking circles, may be such an intermediate solution.[68] Given that pumping trillions of dollars into the financial system in a top-down way by the central banks has not achieved the proclaimed intentions, it now seems justified and necessary to try out new and unconventional approaches such as the bottom-up infusion of money.

11.4   The Way Out: There Is a Better, Alternative Future

It seems that, in the past decades, many people did not pay enough attention to the question why we have such a terribly unsustainable system in the first place. On the one hand, we have been "brainwashed" to think that "everything is ok" (or at least will be). On the other hand, our political and innovation systems have encouraged creative people to innovate within the existing system, but not over the system itself. The orientation at scientific performance indices, the requirement to raise third-party funds, and the institution of peer review encouraged this. Now, we are lacking alternatives and it has become a matter of life or death. However, it is highly unprofessional to fixate ourselves to the system that we have at the moment. There could be many more financial, economic, social and political systems that we have not explored yet—most likely many better ones.

My point is: we should engage in systemic pluralism and should be much more experimental. In fact, with virtual worlds and multi-player online games, we have now the technologies to make large-scale experiments with many different kinds of systems before we implement them. For example, financial systems serve to coordinate the use of scarce resources, i.e. to decide who will get howmuch of what resource. However, there are many different coordination systems that can accomplish this task, and probably many better ones. The current system matches supply and demand on average, but not in detail. As a consequence, we have both, hunger and obesity in the world, which is bad for billions of people. Mechanisms used to decide about organ transplants or to run smart grids are much more sophisticated in matching supply and demand—they do it in a context-dependent and fair way. It is important, however, to ensure that these mechanisms are transparent and fair.

Today's financial system has major flaws. Not only is it prone to devastating cascading effects. It is also essentially one-dimensional. As a consequence, the only possible temporal evolution is to go up or down, which automatically produces

---

[68]Ben S. Bernanke, What tools does the Fed have left? Part 3: Helicopter money, Brookings (11 April 2016), http://www.brookings.edu/blogs/ben-bernanke/posts/2016/04/11-helicopter-money

booms and recessions. Furthermore, it ranks every person and every country on a one-dimensional scale. Therefore, there are automatically winners and losers. This makes the current financial system a control system rather than a system empowering everyone to do business and thrive.

Moreover, since we give more and more weight to economic issues (as I elaborated above), we end up projecting the complexity of the world on one dimension. Money makes the world go round, and all that counts is making money. This is known as utilitarian approach and leads to an oversimplified management of our complex world (often characterized as "linear thinking").[69] It is the one-dimensional optimization implied by today's financial system, which gradually corrupts the functionality of complex dynamical systems such as our society, which then causes major failures. For humans and societies many things matter. Neglecting some of them leads to low economic performance and societal dysfunction. It is well-known that the most diverse economies and societies perform best, not those that have the simplest organization.

In fact, we should understand our financial system, the economy and society as complex dynamical systems, where many components (e.g. people, companies and institutions) adapt to each other.[70] Such systems can only be managed well with multiple different "control variables". To illustrate this, take another complex system: our body. We cannot live just on one thing, say water. We also need oxygen, carbohydrates, different kinds of proteins, vitamins, and minerals. They cannot just be substituted for each other. The shortage of any of them will cause a disease. Healing a disease will require to take medicine, but the right one. More of that medicine will not produce better results, but may poison the body. The right medicine needs to be taken at the right time, and there may be unfavorable interaction effects with other medicines, which need to be considered, too. The same applies to our society. There is not one medicine (money) that can cure all ills. Instead there are many things that matter, but these have been increasingly neglected as compared to money (GDP per capita), and that is the true reason why our economy and society are about to fail.

So, what to do now? We need a multi-dimensional money or incentive system[71]—a system that I will discuss later in connection with capitalism 2.0 and socio-ecological finance.

---

[69]The utilitarian approach measures the value of everything in units of money. It has lead to a one-dimensional optimization, as science and politics and other societal institutions are increasingly influenced by business interests. The related lack of a multi-faceted approach has significantly contributed to the increasing dysfunctionality of many societal institutions, i.e. their difficulty to fix the problems society is faces with. For example, the resulting system does not seem to serve the majority of people well anymore, as elaborated below.

[70]Bertelsmann Stiftung (ed.), To the Man with a Hammer—Augmenting the Policymaker's Toolbox for a Complex World (Bertelsmann, 2016).

[71]Dirk Helbing, Qualified Money—A Better Financial System for the Future (2014), http://papers. ssrn.com/sol3/papers.cfm?abstract_id=2526022 and Dirk Helbing, Interaction Support Processor (2015), https://patentscope.wipo.int/search/en/detail.jsf?docId=WO2015118455

As Albert Einstein said: "We cannot solve a problem within the paradigm that has created it." We need a paradigm shift. In fact, this paradigm shift is already on the way. In the heart of capitalism, we have seen disruptive change. With the invention of Bitcoin, a bottom-up creation of money has come into existence. Moreover, Uber and AirBnb are challenging the classical economy. Surprisingly, the biggest service providers of transportation and accomodation do not own any vehicles or hotels. They just coordinate people and resources more efficiently. With the spread of such sharing economy models, access to services becomes more important than owning material property. A similar thing applies to the music and movie industry, where the business model of streaming seems to replace buying and lending. I, therefore, predict that principles of the sharing economy will be central to our future economy. We should organize it in a more participatory way, however, in the spirit of an open information, innovation, production and service ecosystem.

The exact framework will still have to be sorted out. From my point of view, doing this now is much more important than establishing free trade and service agreements. The last decades have shown that more efficient production and big solutions, which generate the greatest revenue, do not fix the world.[72] We have rather to engage in diverse, mutually complementary solutions, which requires the creation of an information, innovation, production and service ecosystem. As small and medium-sized enterprises are the backbone of thriving economies,[73] it is necessary to revitalize them. Remember that the most diversified economies thrive most.[74] It is also becoming increasingly clear that the civil society can play a much bigger role in addressing the challenges of the future. For example, citizen science[75] and the maker community[76] (using 3D printers and other cheap technologies to produce complex products locally themselves) are important developments in this direction.

Last but not least, young generations appear to be different from previous generations in a number of relevant points: They seem to value a meaningful job more than making a steep career. They have already a larger degree of networked thinking. Friends and family tend to be more important to them than becoming rich. Owning private property is less relevant to them than access to good services. All in

---

[72]The issue with big solutions is that, if it (later) turns out that they have serious drawbacks or side effects, then there is a big problem, potentially a global-scale one. Diversity hedges such risks and ensures that there are alternatives, in case one solution fails.

[73]See footnote 52.

[74]See footnote 33.

[75]Chris Coons, The government wants you to help it do science experiments, Wired (30 September 30),http://www.wired.com/2015/09/government-wants-help-science-experiments/; see alsohttps://www.congress.gov/bill/114th-congress/senate-bill/2113

[76]A Nation of Makers, https://www.whitehouse.gov/nation-of-makers; see also http://www.nationofmakers.org/

all, they are probably adaptable to our future. Moreover, we see new types of companies, which are organized in a bottom-up way and outperform classical, top-down organized competitors.[77] In other words, a new economy, a new society is emerging in front of our eyes. We are about to step into a new era: a digitally empowered, participatory market society.[78] However, as we do not have much time to accomplish this transformation, we must now create a suitable framework for the digital society and make the necessary steps quickly.

11.5 What Can Now be Done

First and foremost, we need paradigm shifts and fundamental change. What short-term action might be taken?

Big Data and Artificial Intelligence (AI): If properly used, these technologies can help to identify inefficiencies and better solutions. However, there are a number of potential pitfalls one needs to pay attention to, otherwise one may produce more harm than good.[79] It is also important to realize that even superintelligent systems will not solve all problems of the world (for example, problems related to friendships, social capital, culture, unemployment or peace).

Water: Given that the production of one portion of meat requires thousands of liters of water,[80] an efficient way of reducing water consumption would be to lower the consumption of meat. In fact, vegetarian and vegan restaurants are increasingly "in". It would certainly help, if there were vegetarian chef shows on TV and every restaurant menu would start with a page with tasty vegetarian dishes. Modern meat replacement products based on soy and other ingredients also seem a promising way to go.

Food: Eating less meat also makes it easier to feed the world population. Growing vegetarian food requires one tenth of the energy required to produce the same

---

[77]Examples are presented in this movie (in German): https://vimeo.com/157724336, https://vimeo.com/157708354

[78]See footnote 34.

[79]See https://www.youtube.com/watch?v=_rfHNvHu8OE, https://www.amazon.com/Thinking-Ahead-Digital-Revolution-Participatory/dp/3319150774 and Helbing, D., van den Hoven, J., Responsible IT innovation: How to digitally upgrade our society?, preprint (2016); Autonomous weapons: An open letter from AI & robotics researchers,http://futureoflife.org/open-letter-autonomous-weapons/; Dirk Helbing, Machine intelligence: Blessing or curse? It depends on us!, Telekom (1 March 2016), https://www.telekom.com/company/digital-responsibility/304108 (reprinted with permission in Chap. 4 of this book).

[80]Ami Sedghi, How much water is needed to produce food and how much do we waste? The Guardian (10 January 2013), http://www.theguardian.com/news/datablog/2013/jan/10/how-much-water-food-production-waste

amount of calories from meat. Further attention must be paid to the regeneration of soil. Suitable planting cycles for food can help. Furthermore, "urban mining" allows one to recycle the scarce resource of "phosphorus".[81] Nitrogen can be extracted from air. Therefore, the production of fertilizer may be sustained for a longer time. Besides, better logistic may make 50% more food available, which is wasted today.[82] Last but not least, urban gardening, local supply chains and local cuisine are becoming new trends.

Reproduction rates: With the spread of Artificial Intelligence and Virtual Reality technology, human reproduction rates will go down for at least two reasons: (1) In a few decades, humans will not anymore be the smartest species on Earth, the "Crown of Creation". Robots can substitute them in some places, and people will adapt to this. (2) Virtual Worlds will become more interesting than our physical world. The current trend is that young people interact with each other more through the Internet than in physical space, which does not produce offspring.[83]

Reduction of "climate gases": Approximately one third of carbon dioxide (CO_2), which seems to be a major contributor to climate change, results from transportation, one third from industrial production and one third from heating. The latter can be reduced by better insulation and by reducing the amount of space heated (or other heating technologies such as infrared heat, which primarily warms up organic matter). Air transport of goods can be increasingly reduced by local production (including 3D printing). Holiday trips by plane may be replaced by local leisure activities. Virtual reality and the new technology of holoportation[84] will, in principle, allow everyone to enjoy any place in the world at home.

Furthermore, car traffic can be reduced dramatically. Google (now Alphabet) is planning to offer "transport as a service" using self-driving cars. This may reduce the number of cars needed to about 15% of the number today, without any loss of mobility. It will also reduce the number of garages and parking lots needed, etc. In other words, the consumption of materials to offer high-quality mobility will drop dramatically. In a further step, we will see a sharing of work space and space for living. Today, we live at home half of the day and we work in another place half of the day; in between we commute and do other things. This is highly inefficient. Robotic furniture will enable reconfigurable space in the sense of working and

---

[81] Andrew Urevig, New study finds recycled Phosphorus could fertilized 100% of U.S. corn, Ensia (15 January 2016), http://ensia.com/notable/new-study-finds-phosphorus-could-fertilize-100-percent-of-u-s-corn/

[82] At least one third of food is wasted today, a maximum of two thirds is used. Therefore, if we can avoid wasting food, the world can eat at least 50% more than today without increasing food production.

[83] "Generation Beziehungsunfähig": Darum macht Tinder süchtig, Huffington Post (7 April 2016), http://www.huffingtonpost.de/2016/04/07/beziehung-tinder-suechtig_n_9632410.html

[84] Holoportation technology is presented in this video: https://www.youtube.com/watch?v=7d59O6cfaM0

living. Virtual reality and holographic[85] or holoportation technology will enable a new level of remote teamwork. Furthermore, coffee shops are starting to offer working environments for business meetings. All in all, this will lead to less commuting, more efficient use of urban space, and attractive, walkable, livable cities.

Methane is another climate gas to care about. Reducing meat production will also lower the emission of this gas.

Energy: Wind power combined with a battery storage system, solar power and geothermal power are most likely our sustainable energy sources of the future. It will be also important to save energy and match supply and demand better. Smart grid technologies, combined with electrical cars (as power storage systems), are making significant progress in this direction.

Altogether we see a tendency towards decentralization in energy production. Since globalization is hitting its limits, this is also found in production (with reindustrialization, 3D printing, etc.). In the financial system decentralization is occurring, too (with BitCoin, crowd funding, etc.). The trend points to more diverse rather than big, "one size fits all" solutions. Cities around the world with similar interests will build coalitions to solve their problems together. We could also organize city olympics, i.e. competitions for the best kinds of technologies, solutions, and city-wide implementations, driven by ambition, competition, engagement and fun.[86]

Innovation: Given that we have not seen enough innovation to solve the impending resource issues, we need to massively improve on responsible innovation. All obstacles have to be kept out of the way.

Academic institutions must be put in the position to work with the data they need and the latest technologies. The equipment of many universities (and schools) today is outdated, and the organizational framework does not keep pace with the speed at which our reality changes. Education needs to get personalized, and the use of Virtual World technologies and gamification can make learning rewarding and easy. Research should be freed from political and economic dependencies. The funding system should be changed from funding of promises to refunding for successful publications and deployments. Interdisciplinary research centers for research on complex systems and Global Systems Science should be massively supported. Talented junior researchers should be able to work on the subject of their interest, get a longer-term perspective and be able to join the research team they like (which should come with an overhead for the receiving institution).

---

[85]The first holographic smartphone is presented here: https://www.youtube.com/watch?v=4tM5qJFsXeM

[86]Dirk Helbing, Countering climate change with climate Olympics (4 January 2014), https://www.youtube.com/watch?v=TaRghSuzBYM

In the digital economy (concerning non-material goods and services) there should be no patents, or they should be opened up for everyone after a maximum period of 2 years.[87] After a 2-year period, software code should also be made open source. Copyrights should be replaced by a simple gratification system. For example, it would be possible to develop a special search engine that compares new files with older ones to determine their degree of similarity. The result could trigger an automatic payment to gratify creators of new ideas, services and products. Such a system would massively promote open innovation.

Patents concerning the material economy should be opened up for a reasonable fee to everyone. It should not be possible to buy a patent or company to take competitive technology from the market.[88] Generally, any unused resource should be opened up for the use by others for a reasonable compensation.

I think these changes are necessary and justified, given that we are talking about the future survival of a lot of people. We also need to engage in mass innovation. Therefore, it is recommended to support citizen science, the maker community, and similar initiatives that are committed to open data, open source, open innovation, etc. For example, an international "Culturepedia" project could identify the success principles underlying the diverse cultures of the world and operationalize them. This would allow people from all over the world to learn from each other and to combine their success strategies in entirely new ways such that better solutions to existing problems would be generated.

11.6   Democracy 2.0 and Capitalism 2.0: The Perfect Couple

By now, it has become clear that we need much better solutions to the world's problems. The challenge is that nobody can fully grasp the complexity of today's world, and that many actions we take have feedback, side or cascading effects. How to come up with better solutions?

We need to put the best knowledge and ideas of the greatest minds (and of artificially intelligent systems) together, i.e. we need to create collective intelligence. Surprisingly, research in the area of collective intelligence shows that the

---

[87]I am aware that this position on patents on copyrights is controversial, but a policy change is overdue and already happening: Tesla has opened up many of its patents. Google and others have open-sourced their Artificial Intelligence software. The recent US court rulings on Google Books supports "fair use" of intellectual property. The trend to open innovation is clearly visible.

[88]Large companies are often weak in terms of innovation (basically, because they are not flexible enough and "new ideas are the enemies of existing ones"). This is the reason why big business tends to buy innovative small and medium-sized enterprises. However, this takes some of the best ideas from the market, and makes them inaccessible to others. Sometimes these innovations are not used at all but just locked away. Such a situation is not in the public interest and does not use resources efficiently. By the way, even though Alphabet (formerly: Google) pursues a phenomenal number of highly ambitious projects in the Google [x] lab, it cannot considered to be a counterexample: over 90% of Google's revenues are made with one single product: personalized information and ads.

combination of several solutions (e.g. a simple average) often performs better than the best individual solution.[89] A precondition for this is that the individual solutions are diverse and independently produced (which speaks against "big nudging", as it tries to align perspectives). In other words, diversity beats the best. The Netflix Challenge, for example, demonstrated this in a very impressive way.[90] All in all, if we replace top-down decisions and majority decisions by a collective intelligence approach, solutions to complex problems will be much better.

This insight calls for a digital upgrade of democracy: democracy 2.0. As I said before, it is not majorities of people that matter for good outcomes, but a suitable combination of diverse ideas. In other words, to improve over today's solutions, we need to take more minority and opposition perspectives on board.[91] If organized well, this does not have to slow down decision-making—on the contrary. Online deliberation platforms now offer the possibility to put all arguments on a virtual table, to organize them in argument maps and identify the different perspectives. Once this has been done, it is time to organize a round table, where the leading representatives of the different perspectives are invited to develop integrated solutions, which bring several perspectives under one roof. In the end, the parliament would select one or a few solutions that satisfy many perspectives. I suggest to choose as many best practice solutions as needed to reach a majority of about two third of all elected representatives.[92] From this set of most promising solutions, the relevant communities (e.g. nations, regions, cities, or spatially distributed interest groups—depending on the level of organization) would then choose the solution, which fits their local needs and culture best. This procedure would create an optimal balance between standardization and diversity.

The performance of the implemented solutions and the relevant success factors would be evaluated from various perspectives. Inferior solutions would be replaced by better ones, and the best solutions would be further improved based on the experience made. This approach would capitalize on creativity, science, and intelligence (including AI), on crowd sourcing, deliberation, and success principles of nature, namely experimentation in niches and selection. The proposed approach combines bottom-up and top-down elements in an innovative way. It is also well

---

[89]Scott E. Page, The Difference: How the Power of Diversity Creates Better Groups, Firms, Schools, and Societies (Princeton University, 2008).

[90]Netflix Prize, https://en.wikipedia.org/wiki/Netflix_Prize; see also the previous reference

[91]It may seem surprising that making "compromises" with minorities would improve the overall system performance. However, the reason is well understandable. In a complex optimization problem, there is typically one solution which is best for a given goal function (perspective). However, there are often many solutions that reach 95% of the best possible performance. Among these solutions, there will be some solutions, which also perform well from other perspectives (goal functions). In other words, when a system is not over-optimized in a single dimension, there is potential to meet many different interests and needs. In this way, the solution will create opportunities for many, which creates large socio-economic benefits.

[92]As of today, the parliament will usually decide for one option only with a 50% majority, but it overstandardizes the world (reduces necessary diversity), which disadvantages a large number of companies and people.

compatible with a federal (region-based) organization and with the well-established subsidiarity principle. Considering the current centrifugal forces in the European Union and the growing polarization in other countries (e.g. between rural and urban areas, or between groups with different subcultures), it becomes increasingly clear that the over-standardized "one size fits all approach" (as decided by the majority or most powerful) does not work anymore. If we do not upgrade democracy by eGovernance, as described above, we are likely to see catastrophic failure, i.e. the fragmentation of political units or social unrest.

An important element in this renovation of our political system is to enable citizens to make better decisions. This requires more access to high-quality information for all and more awareness. With the aim of making progress in this direction, my research teams at ETH Zurich and TU Delft, together with partnering teams at other universities, have started to develop the Nervousnet platform (see nervousnet.info). Our role models are Linux, Wikipedia and Open-StreetMap. The Nervousnet platform allows people, companies and devices to engage in three ways: (1) by generating and contributing data, (2) by analyzing the crowd-sourced datasets, and (3) by sharing code and ideas. In other words, Nervousnet's goal is it to provide real-time data for all and an App-Store for Internet-of-Things applications. Anyone is able to create data-driven services and products using a generic programming interface. The aim is to yield societal benefits, business opportunities and jobs. Nervousnet uses distributed data storage and distributed control, so that it is more robust to attacks and centralized manipulation attempts, easy to scale up, and tolerant to faults. Nervousnet's approach is also compatible with the principles of informational self-determination and, according to our judgment, with the new EU Data Protection Directive.

In particular, Nervousnet will be useful to measure external effects of interactions between people, companies and the environment (so-called "externalities"). Negative externalities (such as noise, pollutants or waste) would get a price, positive ones (such as cooperation, new jobs, or recycling) would get a value. In such a way, a circular economy would be boosted. Furthermore, we can now build a multi-dimensional incentive and exchange system, which I call "socio-ecological finance" (or finance 4.0). This would use knowledge from complexity science and new financial technologies (FinTech) such as Blockchain technologies (as it is used by BitCoin).[93]

The multi-dimensional incentive and exchange system will facilitate the creation of feedback loops in the system. In such a way, it will become possible to support the ability of socioeconomic systems to self-organize in a favorable way. Note that, in economics, such selforganization processes have traditionally been called "invisible hand" phenomena. From complexity science, it is known that self-organization is a natural phenomenon in complex dynamical systems (such as the financial and economic system), but it does not necessarily lead to favorable outcomes (as the financial crisis has shown). However, it is also known that desired structures,

---

[93]See also the chapter on The Blockchain Age (Chap. 13) in this book.

properties or functions will automatically and efficiently occur in case of suitable interactions in the system. The kinds of interactions, which are needed for this, can be determined by complexity theory, computer simulations, laboratory experiments, or multi-player online games. The multidimensional incentive and exchange system mentioned above will allow one to adjust interactions in socio-economic systems such that desired outcomes will result. In other words, 300 years after its invention, it now becomes possible to make the "invisible hand" work, by combining Internet of Things technology with FinTech and complexity science.

The "socio-ecological finance" platform will complement the current financial system and fix its functionality (given that pumping trillions of money into the system in a top-down way could not reach the desired effects). It will allow for the bottom-up creation of multiple new currencies by means of crowd-sourcing activities (namely, by measuring externalities). This approach will enable an effort-based, affordable living without the need of a basic income or helicopter money. It will also make it possible to create taxes as basis for public investments. All of this will not depend on the level of classical employment, i.e. it will be perfectly suited for a highly automated economy.

The concepts of democracy 2.0 (based on online deliberation and collective intelligence) and of capitalism 2.0 (based on socio-ecological finance and sharing economy principles) are well compatible with each other. I consider it the "perfect marriage" between liberalism, capitalism and democracy. This marriage is enabled by novel digital technologies. The socio-economic system built on democracy 2.0 and capitalism 2.0 is organized in a participatory, decentralized and bottom-up way. It offers individual and entrepreneurial freedom. By considering externalities, it supports environmental benefits, coordination and socio-economic order with very little regulation (as compared to today). For the end customer, the consideration of externalities would not be more expensive on the long run—on the contrary. Products and services would be better and cheaper, as long-term costs would be reduced.

The new system would be able to benefit everyone: citizens, politics, and the economy. It would create new opportunities and sources of income for individuals, small and mediumsized enterprises, and big business. A redistribution of money is not implied.[94] The "finance 4.0" system rather uses the fact that non-material products and services, which will increasingly characterize the digital economy of the future, are unlimited. Complementary, sharing and recycling un- and underused material resources will reduce resource shortages. Altogether, this will allow us to create a higher quality of life for more people in the world, if we just slightly adapt our socio-economic framework as outlined above. The proposed systemic changes would be "minimally invasive", but highly effective, and, therefore, their implementation is realistic.

---

[94]In the finance 4.0 system, different kinds of money would be created by crowd sourcing (e.g. by measuring environmental impacts). This money would then "evaporate" and rise up to the top, whereby it passes all levels of society and benefits all of them.

The functional principle of the finance 4.0 system may be illustrated by the following picture: the current digital economy is a little bit like a desert, in which there are just a few palm trees (corresponding to a few big IT companies). The bottom-up creation of money with the socio-ecological finance system will water the desert and add some fertilizer (compare the different currencies with different kinds of minerals added to the soil). This will turn the desert into a rain forest—a highly diverse and interdependent ecosystem (with many different kinds of companies of all sizes), in which there are more than enough opportunities for everyone, and where the biggest trees are certainly bigger than the palm trees in the desert. In other word: the "finance 4.0" system and a suitable framework for the digital economy (requiring, in particular, a sufficient degree of interoperability, exchange, and opening up of currently unused resources for use) can benefit everyone, from the poor to the rich. It is time to build this novel, future-proof system, before the current one (which certainly created many benefits but now reaches its limits) collapses.

11.7   Summary, Discussion, and Outlook

I have argued that the world may soon be confronted with serious resource shortages (e.g. of water, nitrogen, phosphorus,[95] rare earths, and other materials[96]), and that big business and governments have prepared for future crises. These preparations include mass surveillance, censorship and propaganda (by means of personalized information, big nudging, social bots etc.), a behavior-based rationing of resources (potentially using Citizen Scores), armed police, and detainment camps—measures to enforce public order that may appear plausible in a state of emergency. However, I have also shown that the totalitarian systems, which are on the rise, are endangering freedom and diversity and thereby undermining the innovation that we need to master our future.[97] This harms the performance of the economic system and the functionality of society, which can finally lead to economic and societal collapse, as it is unfortunately quite likely now.

I have also pointed out that, due to the digital revolution, business models and institutions are undergoing fundamental transformation. The digital revolution allows us to reinvent everything, and we will therefore soon live in a differently organized, digital society. This future will be based on networked thinking and collective intelligence. The paradigm of selfish optimization (also known under the label "homo economicus") will not anymore be competitive with collaborative

---

[95]Will Steffen et al., Planetary boundaries: Guiding human development on a changing planet. Science   347   (2015).   http://science.sciencemag.org/content/347/6223/1259855.full-text.pdf +html, p. 736.

[96]See footnote 39.

[97]Note that most innovations happen in a bottom-up way, and many of them "by accident".

models of the future,[98] which engage in partnerships of companies, suppliers and customers, users, citizens, and patients. The novel, partner-based approach is often characterized by terms like "co-creation" or "information, innovation, production and service ecosystem".

The future digital society is able to benefit everyone, as the resources in the emerging digital economy are basically unlimited. Due to automation (Artificial Intelligence, Robotics), the old economy making money by rationalization (using "economies of scale") will be run with about 50% of today's manpower (if the numbers that 50% of today's jobs will soon be gone[99] are correct). The other 50% will eventually produce digital products and services in the new, non-material economy, which is now emerging. This new, digital economy encompasses, in particular, Virtual Worlds, which will be used as "experimental worlds" and for new kinds of business. In other words, in the long run I do not expect mass unemployment, but we may face intermediate challenges during the transformation process. This calls for answers bridging from the old to the new world (such as helicopter money—until the socio-ecological finance system is fully operational). Without such bridging solutions, we may, in fact, run into revolutions or wars. However, by now, we should have learned enough from history to avoid such disasters and to avoid the repetition of old mistakes.

Besides the challenges we are faced with, I have sketched concrete solution approaches and, in particular, an organizational framework of the future digital economy and society: democracy 2.0 and capitalism 2.0. These will upgrade today's democracy and capitalism, and be married together by means of a combination of new technology and science: the Internet of Things, FinTech, and complexity science. Democracy 2.0 (using online deliberation platforms) and capitalism 2.0 (crowd-sourcing various kinds of money by measuring externalities in order to run a highly differentiated incentive system) are approaches complementing superintelligent systems.[100] While superintelligent systems aim at the top-down optimization of systems, democracy 2.0 and capitalism 2.0 work in a bottom-up way. They support innovation and (co-) evolution as well as the interaction, self-organization and coordination of diverse activities and interests, of people and resources. Future, open and participatory sharing economy platforms are part of this solution.

Why are open, participatory approaches so important? Because they enable a better, more innovative use of scarce resources. I claim that not material resources are our main bottleneck, but the way we are using them. Let me illustrate this with an

---

[98]Dirk Helbing, Economics 2.0: The natural step towards a self-regulating, participatory market society, Evol. Inst. Econ. Rev. 10(1) (2013), pp. 3–41.

[99]Carl B. Frey/Michael A. Osborne, The future of employment: How susceptible are jobs to computerisation?, Oxford Martin (2013), http://www.oxfordmartin.ox.ac.uk/downloads/academic/The_Future_of_Employment.pdf

[100]The usefulness of superintelligent algorithms is limited not only by the data available, but also by the complexity of the system. Complex dynamical systems that need high levels of innovation, such as our economy and society, require decentralized bottom-up approaches to perform well.

example from traffic flow management in cities, where we have made a remarkable discovery: When we replaced the attempt of optimal top-down control of traffic lights by a bottom-up approach based on principles of self-organization,[101] we achieved a 30–40% higher performance.[102] How is this possible? Complex optimization problems cannot be solved in real-time, even with supercomputers. One needs to make simplifications, and the optimization is, therefore, performed in a certain, simplified solution space (for example, periodically operated traffic lights). However, some of the best solutions lie outside of the solution space chosen, no matter what solution space one may choose.

In contrast, the self-organization approach attempts a flexible adaptation to the actual local needs. In our approach,[103] traffic flows control the traffic lights rather the other way round. This is based on simple rules that promote the coordination of neighboring traffic lights. In other words: we do not limit the solution space—we let the system evolve according to its needs and encourage coordination by the kinds of interactions implemented between the traffic flows and the neighboring traffic lights.

Why is this of general importance? In urban traffic systems, road capacity (storage space) and flow capacity are often short resources, and their management is critical, if massive traffic jams shall be avoided. In the economy, materials are the resources that may run short. Suppose we get into a situation where we can use only 20% less resources than usual. So, assume that everyone would get just 80% of the resources that were available before. Then, one may never have enough resources to produce a particular product, because certain parts or ingredients would be missing. In other words, a proportional rationing would trigger second-order shortages and make the situation worse than it would have to be. Instead, we need a flexible system, where everyone can get 100% of certain resources on some days, but 70% on others, to compensate for this. This corresponds to a turn-taking principle[104] (which actually bears some similarity with the coordinated switching of traffic lights).

In essence, a flexible system is able to create more goods and services than one that is managed in a top-down way.[105] Today, we know how to build such flexible

---

[101] Stefan Lämmer/Dirk Helbing, Self-control of traffic lights and vehicle flows in urban road networks, J. Stat. Mech. P04019 (2008); Stefan Lämmer/Reik Donner/Dirk Helbing, Anticipative control of switched queueing systems, Eur. Phys. J. B 63 (2008), pp. 341–347.

[102] Stefan Lämmer/Dirk Helbing, Self-stabilizing decentralized signal control of realistic, saturated network traffic (2010), http://www.santafe.edu/media/workingpapers/10-09-019.pdf; further publications are available here: http://stefanlaemmer.de/?content=Publikationen; for internal reports contact dhelbing@ethz.ch or traffic@stefanlaemmer.de

[103] See footnotes 101 and 102.

[104] Dirk Helbing et al., How individuals learn to take turns: Emergence of alternating cooperation in a congestion game and the prisoner's dilemma, Advances in Complex Systems 8(2005), pp. 87–116.

[105] This is known from the competition between capitalism 1.0 and communism 1.0. Note that, in times of scarcity and rationing of resources, black markets emerge, which try to compensate for a lack of flexibility.

systems. In connection with congestion and route choice problems, for example, I have elaborated concepts that can increase the traffic capacity considerably, based on principles of fairness and flexibility.[106] In connection with gas supply, we have furthermore elaborated, how fairness can be reached, based on decentralized approaches.[107] Interestingly, symmetrical interactions tend to support not only fairness, but also optimal self-organization.[108]

Today, we still lose about 30% or more of many perishable goods (such as food) due to bad logistics. Improving logistics along the lines discussed above would allow one to set many of the currently unused resources free.[109] Rationing resources and applying citizen scores, in contrast, would reduce the amount of flexibility, and let a socio-economic system perform badly.

Big Data and Artificial Intelligence are certainly useful tools to identify inefficiencies and underutilized solutions. However, like any other tool, they are not solutions for everything. We must also see their limitations and side effects. The entirely new solutions and paradigms that are now needed to master our future will not be delivered just by brute-force data mining or machine learning. We must further see the serious dual use problem of powerful digital technologies. As they have societal-scale impact, they can potentially cause much greater damage than a nuclear meltdown or atomic bomb (which is geographically limited). Think, for example, of a blackout of the Internet for a couple of days,[110] or even for months, as it might happen as a result of solar storms ("space weather").[111]

Powerful tools, particularly those with a centralized architecture, imply the risk of random or triggered failure, and of serious misuse. They are attractive for organized criminals, terrorists and people with extreme agendas. We therefore need to be wary of people or institutions who may instrumentalize future crises to turn democracy

---

[106]Dirk Helbing, Dynamic decision behavior and optimal guidance through information services: Models and experiments, in: Michael Schreckenberg/Reinhard Selten (eds.), Human Behaviour and Traffic Networks (Springer, 2004).

[107]Rui Carvalho et al., Resilience of natural gas networks during conflicts, crises and disruptions. PLoS ONE 9(3): e90265 (2014).

[108]Dirk Helbing/Tamás Vicsek, Optimal self-organization, New Journal of Physics 1 (1999), 13.1–13.17.

[109]Dirk Helbing/Stefan Lämmer, Method for coordination of competing processes or for control of the transport of mobile units within a network, https://www.google.com/patents/US8103434

[110]Rod Beckstorm, What if a hacker caused a large-scale Internet outage, WEF (12 June 2012), https://www.weforum.org/agenda/2012/06/what-if-a-hacker-caused-a-large-scale-internet-outage/; Matthias Schüssler, Die unterschätzte Gefahr eines Internetblackout, TagesAnzeiger (12 April 2016), http://www.tagesanzeiger.ch/digital/internet/Die-unterschaetzte-Gefahr-eines-Internetblackout/story/30794811

[111]Sonnenforscher warnen vor dem "Big One", Der Bund (8 April 2016), http://www.derbund.ch/wissen/natur/sonnenforscher-warnen-vor-dem-big-one/story/25117391; Geoffry Reeves, The space weather threat…and how to protect ourselves, HUFFPOST Science (19 April 2016), http://www.huffingtonpost.com/lab-notes/theunpredictability-of-s_b_9721612.html

into fascism 2.0, communism 2.0, or feudalism 2.0.[112] To minimize potential misuse by mistake or intention, we must create a suitable framework, which ensures security, accountability, and transparency; democratic control, scientific, responsible, ethical and pluralistic use; forgetting, privacy and informational selfdetermination. One should also aim at openness (open data, open source, open innovation) and participatory opportunities to enable co-creation and benefits for everyone (as much as justified and possible). Altogether, the goal should be to create an information, innovation, production and service ecosystem for people, ideas, resources and initiatives, including an open sharing economy platform. The Nervousnet platform mentioned above could contribute to this ecosystem.

Last but not least, it must be realized that centralized information systems may be unsustainable on the long run. Cybercrime causes a damage of around 1 trillion dollars per year and is exponentially increasing.[113] Adding more devices to the Internet increases its vulnerability. CIA chief Clapper considers the Internet of Things to be the greatest threat to the USA.[114] In fact, there is no 100% security anywhere. The US military has been hacked, the Pentagon, the White House, the German Bundestag, and probably every company, too. The idea that systems (particularly learning ones) can be made 100% secure if we just control them more and more is a dangerous illusion—and tends to end in a loss of freedom and security.[115] Therefore, we may need a new security paradigm oriented at resilience, which could be inspired by the human immune system: even though this is attacked by bacteria millions of times every day, we live more than 70 years on average. Remarkably, the immune system is decentrally organized, which ensures that there is no single point of failure. For this and other reasons, the Nervousnet platform will be a decentralized system. Error and attack tolerance have been guiding principles since the creation of the Internet—we should not forget about this.

---

[112]In this connection, determining who financially supports populist movements is strongly advised.

[113]Elinor Mills, Study: Cybercrime cost firms $1 trillion globally, CNET (29 January 2009), http://www.cnet.com/news/study-cybercrime-cost-firms-1-trillion-globally/; Jana Rooheart, Cyber crime to reach $2 trillion by 2019, business.com (19 April 2016), http://www.business.com/internet-security/cyber-crime-to-reach-2-trillion-by-2019-what-can-we-do/

[114]Kelsey D. Atherton, Clapper: America's greatest threat is the Internet of Things, Popular Science (9 February 2016), http://www.popsci.com/clapper-americas-greatest-threat-is-internet-things

[115]Benjamin Franklin: "Those who surrender freedom for security will not have, nor do they deserve, either one." http://www.goodreads.com/quotes/140634-those-who-surrender-freedom-for-security-will-not-have-nor

11.8 Heading Towards the Illuminated Age

As the experience in the past decades has shown, despite great achievements, technology alone cannot solve our problems. Human behavior and civil society need to be part of the solution. In order to make quick progress, I am summarizing below some of the steps that can be taken within the next few years:

- Business models and policies that do not comply with human dignity and human rights should be banned.
- Underused resources should be opened up for use for a reasonable compensation. For example, sharing economy platforms can offer opportunities to improve the use of underused resources. Furthermore, new solutions should be pursued to reward people and companies for innovative solutions and creative products.
- A centrally managed one-size-fits-all approach will not be diverse enough to create the innovation rates, collective intelligence and societal resilience needed. One should, therefore, support international interdisciplinary initiatives developing new solutions to impending crises, based on both, competition and collaboration.[116] In this connection, a Culturepedia project, climate olympics and similarly engaging formats might be fruitful.
- One should support the ability of people to help themselves and each other. Big Data, Artificial Intelligence, innovation and production should be democratized by supporting open data and open innovation. Furthermore, one should catalyze an information, innovation, production and service ecosystem by requiring or rewarding interoperability. Then, SMEs, NGOs, citizen scientists and the maker community could better contribute to solving the problems of the world, based on the principle of glocality ("think global, act local").
- One should build "democracy 2.0" ("digital democracy"), which requires suitable platforms for online deliberation and eGovernance. Such platforms can now be

---

[116]Social and cultural diversity is as important as biodiversity for human survival. It hedges risks with respect to unexpected events, which will surely happen at a high rate as we undergo the transformation from the service society to the digital society and from the carbon-based economy to a low-carbon economy. How to reduce conflict under these stressful conditions in a highly diverse world? The idea is to familiarize people with other points of view (which corresponds to breaking "filter bubbles"). The relevant scientific experiment was performed by Muzafer Sherif, see https://en.wikipedia.org/wiki/Realistic_conflict_theory#Robbers_cave_study. In essence, it has been found that big challenges requiring team work can overcome tensions between different groups, see also https://www.youtube.com/watch?v=37QvponcEDc from minute 13. Real-world success stories of this approach are the European exchange program between cities and for students (Erasmus), which have managed to overcome post-war sentiments and establish a basis for peace in Europe. Now, this approach can be scaled up to global scale. Using Virtual Reality (VR), one can enable people to put themselves into other peoples" shoes and understand their perspectives. VR can help to overcome cultural barriers and support international collaborative projects. The advantage of this approach is that it can create social cohesion across national boundaries while allowing for diverse approaches. In a sense, Virtual Reality technology, if used in this way, may be seen as a tool to speed up the evolution of other-regarding preferences, i.e. to turn homo economicus into homo socialis, who has the ability to consider the points of views of others.

created. They can help to bring the best knowledge and ideas together, which is key to master the challenges of the future. (They may also integrate Artificial Intelligence technology.) "Pluralistic" solutions that are acceptable from diverse perspectives have the advantage that they can serve multiple purposes and functions. The parliament could decide for, say, one, two or three of such solutions, giving communities a choice to implement a locally and culturally fitting solution.

This approach would reach a good balance between standardization and diversity.

- Last but not least, one should build "capitalism 2.0" by adding a "socio-ecological finance" system to our current financial system, where various new currencies measure social and environmental impacts and attribute a certain value or cost to them. This socio-ecological finance system would provide a new, multi-dimensional incentive and exchange system, enabling to support the self-organization and coordination in complex dynamical systems such as our economy and society. If suitably specified, socio-ecological finance will foster a circular and sharing economy, which implies the more efficient use of scarce resources and a good quality of life for many people. The system will allow everyone to earn money by crowdsourcing (e.g. by measuring environmental impacts). Taxes for public investments can be automatically created as well. Capitalism 2.0 offers many dimensions to "do well". It also considers non-material value. Finally, as the approach takes externalities into account, it can effectively support environmental care and social cooperation (including peace between cultures) while supporting individual and entrepreneurial freedom.

Until we have created the public framework of the participatory market society to come, we may go through a period of troubled waters. However, we should see it as a time of adventure and discovery, a time where individuals and groups can change the world to the better more than it has been possible in the past. The resulting data-driven society will not be run like a giant machine or clock tower, but rather like a well-coordinated system of diverse and largely autonomous, self-organizing systems, activities and processes. Among the data-driven societies discussed above, the combination of democracy 2.0 and capitalism 2.0 is certainly the most innovative and efficient socio-economic system, and the best possible basis for a thriving society.

In the era to come, material goods will be more efficiently produced, thanks to Big Data, Artificial Intelligence and robotics. Principles of the circular and sharing economy ("reduce, reuse, recycle") will mitigate or even overcome upcoming resource shortages. Ownership will become less important than opportunities to use products and services. Today's mobility may be provided by less than 20% of today's vehicles, and 50% of today's buildings would be enough to provide enough space for work and life. All in all, cities will probably become more livable than today.

Besides the old, material economy, we will see the growth of a new, digital economy. This will be more or less unlimited, as it is immaterial, and will allow everyone—from individuals to big corporations—to benefit, if we get the framework right. We can see already that information plays an ever more important role in our lives (when measured in number of hours spent with information systems). Creative products and ideas will become increasingly important. This also implies that human values will be key again. It is clearly visible that people are currently seeking for new ways of leading meaningful lives. Soon, we will see a new Zeitgeist. The consumption of immaterial goods and services will increasingly replace the consumption of material ones, and consumption will become more active, as it is reflected by the terms "co-creation" and "prosumer" (co-producing consumer).

Furthermore, rather than maximizing gross domestic product (GDP) per capita, many nations around the globe may maximize happiness, soon. The "economics of happiness"[117] shows that people actually do not need a lot of material resources to be happy: water, shelter, information access, and a meaningful life. The latter requires a society built on values and trust; it requires friendships, opportunities for personal self-development, a good health system, and social security.

What kind of values may be guiding us in the densely connected, digital society of the future? Perhaps the following ones, which are the outcome of extensive discussions I had with many people[118]:

1. Respect: Treat all forms of life respectfully; protect and promote their (mental, psychological and physical) well-being.
2. Diversity and non-discrimination: Support socio-economic diversity and pluralism (also by the ways in which Information and Communications Technologies are designed and operated). Counter discrimination and repression, prioritize rewards over punishment.
3. Freedom: Support the principle of informational self-determination; respect creative freedom (opportunities for individual development) and the freedom of non-intimidating expression.
4. Participatory opportunities: Enable self-determined decisions, offer participatory opportunities and a choice of good options. Ensure to properly balance the interests of all relevant (affected) stakeholders, particularly political and business interests, and those of citizens.
5. Self-organization: Create a framework to support flexible, decentralized, self-organized adaptation, e.g. by using suitable reputation systems.

---

[117]Bruno S. Frey, Happiness: A Revolution in Economics (MIT Press, 2010); Rudolf Hermann, Glücksforschung: Die dänische Theorie des Glücks, NZZ (2 April 2016), http://www.nzz.ch/lebensart/gesellschaft/die-daenischetheorie-des-gluecks-1.18720975. Evidence of the psychological literature, specifically self-determination theory (SDT), implies that the following factors largely contribute to happiness: social inclusion, competence, and experience of autonomy; for a related review see Richard M. Ryan/Edward L. Deci, On Happiness and Human Potentials: A Review of Research on Hedonic and Eudaimonic Well-Being, Annual Review of Psychology 52, 141–166 (2001), http://www.annualreviews.org/doi/full/10.1146/annurev.psych.52.1.141

[118]See footnote 34.

6. Responsibility: Commit yourself to timely, responsible and sustainable actions (or omissions), by considering their externalities.
7. Quality and awareness: Commit yourself to honest, high-quality information and good practices and standards; support transparency and awareness.
8. Fairness: Reduce negative externalities that are directly or indirectly caused by your own decisions and actions, and fully compensate the disadvantaged parties (in other words: "pay your bill"); reward others in a fair way for positive externalities.
9. Protection: Protect others from harm, damage, and exploitation; refrain from aggressive or war-like activities (including cybercrime, cyberwar, and misuse of information).
10. Resilience: Reduce the vulnerability of systems and increase their resilience (e.g. through decentralization, self-organization and diversity).
11. Sustainability: Promote sustainable systems and long-term societal benefits; increase systemic benefits.
12. Compliance: Engage in protecting and complying with these fundamental principles.

To summarize the above even more briefly, the most important rule is to increase positive externalities, reduce negative ones, and ensure fair compensation. This might be considered as an operationalization of the golden rule: Behave in such a way, as you would expect it from others, if affected by that decision (where "others" also includes the environment and ecosystem around us). Shall we give it a try?

Chapter 12
How to Make Democracy Work in the Digital Age

Dirk Helbing and Stefan Klauser

Recently, we have heard many complaints about how democracy works these days—or maybe rather why it doesn't work. In a recent Huffington post article, Dhruva Jaishankar, a Fellow at the Brookings Institution in India, claimed that digital democracy is the evil that makes our world ungovernable.[1] We argue that Iaishankar defines digital democracy in a flawed and misleading way. This could cause serious misunderstandings of what the problems are and what are the possible solutions. In the following we will show that digital democracy—if properly understood[2]—is the most promising way to build prosperous societies in the digital age.

This article by Dirk Helbing and **Stefan Klauser** was first published as OpEd in the Huffington Post under the URL http://www.huffingtonpost.com/entry/how-to-make-democracy-work-in-the-digital-age_us_57a2f488e4b0456cb7e17e0f and appeared in modified form in the German book "Smartphone-Demokratie", edited by Adrienne Fichter (reproduction with permission of NZZ Libro).

[1]http://www.huffingtonpost.in/dhruva-jaishankar/brexit-the-first-major-ca_b_10695964.html
[2]D. Helbing and E. Pournaras, Build digital democracy, Nature 527, 33–34 (2015): http://bit.ly/1WCSzi4

D. Helbing (✉)
ETH Zurich, Zürich, Switzerland

TU Delft, Delft, Netherlands

Complexity Science Hub, Vienna, Austria
e-mail: dhelbing@ethz.ch

S. Klauser (✉)
ETH Zurich, Zürich, Switzerland
e-mail: stefan.klauser@gess.ethz.ch

© Springer International Publishing AG, part of Springer Nature 2019
D. Helbing (ed.), *Towards Digital Enlightenment*,
https://doi.org/10.1007/978-3-319-90869-4_12

12.1   Brexit, Trump, AfD: Is the Internet Creating Protest Voters?

Most commenters agree that we see a polarization of society in recent times. This has to do with the way modern mass media and social media work. They tend to create 'filter bubbles' reinforcing the own opinion, while reducing the ability to handle different points of view. They become increasingly personalized, manipulative, and deceptive, spreading oversimplified messages or misinformation.

It is in fact true that the Median Voter Theorem, stating that in a two party system parties should move towards the center,[3] has come to its limits. The political positions of parties in a majority-based two party system do not always develop like this. Sometimes they actually get more polarized. The presidential elections in the US, where candidates tend to stress ever more extreme positions, provide clear evidence of that. In Europe though, we find a different pattern. The centralized bureaucracy in Brussels increasingly supersedes the federal (region-based) organization and, thereby, violates the well-established subsidiarity principle. As the so-called Bryce's Law[4] predicts for young institutions, there is evidence for an increasing centralization of the EU.[5] However, considering the growing diversity in the European Union (with marked differences between rural and urban areas, and between groups of different age or cultural background), it becomes clear that an over-standardized "one size fits all approach" decided by the majority or most powerful does not work anymore. This incumbent-dominated governance engenders unpredictable reactions from diverse citizens, and the evidence shows that the neglected citizens react by voting for protest candidates and/or supporting radical solutions to complex challenges (see Brexit).[6]

12.2   Politics Is Not One-dimensional

It is certainly true that the societal and political complexity has dramatically increased with the multitude of interdependencies in our highly networked, globalized world. Therefore, the question is how to deal with the compounded complexity

---

[3]see also: Hotelling, Harold (1929). "Stability in Competition". The Economic Journal 39: 41–57.

[4]The Bryce's Law describes the tendency of a federal state to become more centralized over time. See: H. Badinger/V. Nitsch (Ed.) (2016): Routledge Handbook of the Economics of European Integration.

[5]L. Olai, L. Lehmkuhl (2012): Centralizing the EU? An analysis of the European Court of Justice's tendency to rule in favor of centralization of the European Union.

[6]Dixit & Weibull (2007) present in their book "Political polarization" a model highlighting that different priors are necessasry conditions for polarization. In the context of this article here, filter bubbles and selective media engender heterogeneous priors amongst the population. Then, when a population is presented with the same information, people arrive at dichotomized conclusions, and politlical opinions shift toward extreme—and not toward the center.

and unpredictable outcomes—even for experts and for expert systems (based on big data and artificial intelligence). Some commenters suggest that voters cannot handle this complexity—they would be easily manipulable and lean towards populist and inadequate solutions. And therefore, voters should be disregarded overall in favor of a data-driven society, they say. Then it would be Google's "omniscient algorithm" or IBM's cognitive computer, called Watson, deciding about what had to be done. Depending on the implementation, this experiment may very well end in fascism 2.0 (a big brother and brave new world society), communism 2.0 (distributing rights and resources based on a "benevolent dictator" approach), or feudalism 2.0 (based on a few monopolies and a new kind of caste system). However, the data-driven variants of governance models that have failed in the past will not suddenly become more acceptable. The question is rather how to use the digital opportunities of today and upgrade democracy, "the worst form of government, except for all the others", as Churchill framed it.

12.3 Democracy 2.0: How to Harness Collective Intelligence by Digital Means

The long-term consequences of centralized top-down control could be devastating due to a loss of socio-economic diversity and resilience, a decline in the innovation rate and socio-economic progress, political instability and war or revolution. Centralized top-down optimization may be a proper paradigm for companies or supply chains, but complex societies need pluralism and combinatorial innovation to thrive. The success principles of the past—globalization, optimization, and administration—have more or less hit their limit. To reach the next level of society, an economy dominated by networks must build on the principles of co-creation, co-evolution, and collective intelligence.

Overall, to achieve culturally fitting, sustainable and legitimate results that leverage the benefits of complexity and diversity, it is crucial to move from a government paradigm based on power to a paradigm based on empowerment. Combining smart technologies with smart citizens is the recipe to create smarter societies. This can be reached by creating Massive Open Online Deliberation Platforms (MOODs), which allow all interest groups to put their arguments on a particular subject on a virtual table, where they can be structured into different points of view. Or in the words of Landa and Meirowitz: "In revealing correct, fuller, or simply better organized information, deliberation provides an opportunity for participants to arrive at more considered judgments themselves and to affect collective decision making by influencing the judgments of others."[7]

---

[7]D. Landa/A. Meirowitz (2009): "Game Theory, Information, and Deliberative Democracy", in American Journal of Political Science, Volume 53, Issue 2, pages 427–444.

In a second step, it is important to work out innovative solutions that integrate several perspectives and, thereby, benefit several interest groups well, not just the incumbent or 51% majority. This is the essence of "digital democracy". It is based on "collective intelligence"—on bringing the knowledge and ideas of many minds (and artificially intelligent systems) together. It is the combination of ideas and interaction of humans that have shown to deliver the best results in most challenges.

While ensuring collective innovation, an updated democratic process should be able to reach equally distributed opportunities and satisfaction, as much as this can be done. While this cannot always be achieved in each single decision, we could certainly get much better in satisfying diverse interest groups than today. Putting it differently, digital democracy is about creating the digital tools to make deliberative democracy as described by a Habermas or Fishkin work efficiently.

So, instead of trying to revive governance principles of the past, which have failed to embrace the complexity and diversity of modern societies, we should engage in digitally upgrading democracy. After all, being the result of many wars and revolutions—democracy is a highly advanced governance system that has taken on board the wisdom of some of the smartest and most respected people in human history. Rather than accepting data-driven governance to control and abate societal diversity and complexity, we propose a way to leverage complexity for our benefit, through a platform of participation and decentralization.

12.4   Overcome the Dictatorship of the Majority

Some scholars state that, especially in polarized societies and societies with significant minority groups, simple majority voting is not adequate, because it can lead to a dictatorship of the majority over the minorities. Additionally, extremely close votes (see Brexit vote in the UK and the vote on mass immigration in Switzerland) often evoke protests about the fairness of this process. The loosing side, representing a significant share of the population, fears suffering under the new rule and has a strong interest to fight for an agreeable implementation of the initiative.

The fact that large minorities are often being ignored presents a serious issue, which is not easy to solve. However, with the means of MOODs, one can find solutions that consider various views on certain aspects of a topic. Today, one of the main problems is that people can only cast a "yes" or a "no", i.e. to either agree on a proposed solution or disagree. The topic is often extremely complex and has many facets. So, letting people decide about "yes" or "no" is simply not enough. We suggest that citizens should be able to continuously engage in a specific type of online deliberation processes, where they can feed in their ideas and voice their preferences on different aspects of a topic. Brexit, for example, has many implications for Britain. Most voters saw some advantages and some disadvantages of it. Had one known how the electorate ranked the importance of the different facets of the political issue (for example, regarding immigration on the one hand and economic interdependence on the other hand), policymakers could have tried to disentangle some of the aspects and prevent severe dissatisfaction of the electorate. One could argue that politics is not a self-service wonderland, so voters would ultimately have to choose between different packages. Nevertheless, it is pretty clear that understanding voters' preferences could improve political processes and lead to tailor-made solutions, especially when a reasonable level of federalism is embraced.[8]

A refined, more inclusive process has several advantages. It enables people to learn about the different aspects of a complex political topic. At the same time they can contribute to the solution from the beginning, which is believed to lead to a higher satisfaction.[9] Analogously, it should diminish the chances that protest movements and extreme solutions will find good breeding grounds.

Even if the results of the deliberation process would not be binding for policymakers, they MOODs would give them ample guidance when drafting new laws. It would also be possible to take regional, ethnic and religious differences into account, which could lead to culturally fitting law-making and easily show whether it makes more sense for a specific law to be adopted on a federal or on a regional level. There could still be a majority vote at the end of a deliberation process. But at this point, the solution would already include a substantial amount of the ideas and wishes of the citizens and it is likely that we would not see extremely polarized situations anymore.[10] Regardless of whether a proposed new law engendered a 50:50 polarization of society, deliberation processes after the vote could substantially lower the dissatisfaction of the minority, especially again, when mixed with a high level of regional autonomy in the way the vote/law is being implemented.

12.5   Counterbalance Misinformation/Spread of Extreme Views Through (Social) Media

Missing media competence, scarce time, filter bubbles and financial means of potent influencers all lead to the spread of inaccurate, misleading or even wrong information. For citizens it is increasingly hard to judge, which information can be trusted and why. Governments, companies and rich individuals today can buy armies of bloggers and social media experts to run profiles and chat bots, which are flooding social media channels with the information they want to spread. Most of today's largest social media platforms have currently no means to moderate these discussions. This information asymmetry contradicts the notion of an authentic

---

[8]The Partido de la Red movement in Argentina makes digital democracy to the central element in its program. See https://www.ted.com/talks/pia_mancini_how_to_upgrade_democracy_for_the_inter net_era?language=en

[9]Compare E. Ostrom (1990): Governing the Commons.

[10]Compare: James S. Fishkin/ Robert C. Luskin (1999): Bringing Deliberation to The Democratic Dialogue.

deliberation process as e.g. defined by John S. Dryzek.[11] We thus need to create new platforms allowing for an informed, balanced, conscientious, substantive and comprehensive deliberation processes.

12.6 Deliberation and Influence Exercised Through Ranking, Voting and Discussions

There are some important features that the MOODs need to have: (1) They should be transparent in the way they work and decentralized, to reduce manipulation and censorship. (2) They should be moderated (by community-elected moderators) to ensure that discussion are constructive and fair. (3) Artificial Intelligence should be used to detect abnormal activity and reveal chat bots as well as ghostwriters. AI could also be used to organize the arguments made and support a multi-faceted discussion. (4) Reputation systems should incentivize responsible behavior and high-quality contributions. For example, one could give a greater visibility to contributors with a higher reputation. It is also important to detect manipulative rating activities. (5) Finally, a transparent and fair qualification mechanism could determine the roles that individuals can play in the deliberation process, and what additional data and platform functionality would be at their disposal.

To define Digital Democracy merely as democratic processes in a media-dominated and digitalized world falls short of what a reasonably advanced idea of Digital Democracy encompasses. A sophisticated model of Digital Democracy must be based on the concepts of co-creation, co-evolution and collective intelligence, enabled through the use of modern digital means.[12]

It certainly takes a substantial amount of work to build these platforms and to upgrade democratic processes to be fit for the digital age. The task we have to accomplish has technical, legal and motivational aspects. Especially the question of how to engage enough people in the deliberation process will be crucial. One has to secure easy access to the platform and experiment with incentives and gamification to reach sufficiently broad participation. However, there are no obstacles that could not be overcome. The potential benefits of a suitably refined (direct) democratic process clearly outweigh the costs of turning history back and neglecting "we the people". Digital democracy supports society's historical achievements: self-determination and freedom, the division of power and fairness, social inclusion and participation as well as diversity and resilience.

---

[11]Dryzek (1990): Discursive Democracy: Politics, Policy, and Political Science.

[12]D. Helbing, Society 4.0: Upgrading society, but how? https://www.researchgate.net/publication/304352735; D. Helbing, Why we need democracy 2.0 and capitalism 2.0 to survive, Jusletter IT (May 25, 2016), see http://bit.ly/1O5axWZ

Chapter 13
The Blockchain Age: Awareness, Empowerment and Coordination

Jeroen van den Hoven, Johan Pouwelse, Dirk Helbing, and Stefan Klauser

Blockchain technology may be the basis of the next step in human, social, and cultural evolution.

Currently there's a lot of hype surrounding blockchain technology. But the best ways to use it are still to come. Blockchain is often seen as a revolutionary technology, a public decentralized registry that allows for trusted peer-to-peer transactions without middlemen such as banks or other institutions. Blockchain technology is used for new kinds of money and payment systems such as Bitcoin and Ether. However, it also enables to create distributed autonomous organizations (DAOs).

Besides the financial sector, blockchains may revolutionize supply chains, the health system, administrations, humanitarian aid and law enforcement. Like any other technology, however, one must pay attention to possible side effects and ethical implications.[1] For example, if you are late paying the interest rates of your

---

This article by **Jeroen van den Hoven, Johan Pouwelse**, Dirk Helbing and **Stefan Klauser** has first been published as ResearchGate draft under this URL https://www.researchgate.net/publication/317239062

[1]C. Dierksmeier and P. Seele, Cryptocurrencies and business ethics, J. Bus. Ethics (2016) https://link.springer.com/article/10.1007/s10551-016-3298-0

J. van den Hoven · J. Pouwelse (✉)
TU Delft, Delft, Netherlands
e-mail: m.j.vandenhoven@tudelft.nl

D. Helbing (✉)
ETH Zurich, Zürich, Switzerland

TU Delft, Delft, Netherlands

Complexity Science Hub, Vienna, Austria
e-mail: dhelbing@ethz.ch

S. Klauser
ETH Zurich, Zürich, Switzerland
e-mail: stefan.klauser@gess.ethz.ch

loan, you may not be able to rent a car, or your access to other services might be blocked.

In an over-regulated world, strict law enforcement might even make our economy and society inefficient and dysfunctional. Our old world used to be a world where it was possible to do things that were morally undesirable. With blockchain we are moving to a world where the morally undesirable is made impossible. Even though this may sound good at first, it may actually prevent learning from mistakes and, furthermore, seriously obstruct innovation—since innovation always challenges established solutions.

A further concern is the tendency that non-commercial content in the Internet may gradually be crowded out. Before we are able to get creative, we may then have to deal with a lot of intellectual property rights. To illustrate the implications, just imagine how ineffective it would be, if we had smart contracts for use of language and, therefore, had to pay for every word we use, when communicating with other people. This would be the end of shared culture as we know it.

Over-commercialization and loss of creative freedoms are, therefore, seriously issues to be considered.[2] This is particularly important in times where automation is forcing us to be more creative, and access to data is very limited for ordinary people, start-ups, small and medium-size businesses. According to the WEF and OECD, wealth inequality is already a serious obstacle to economic growth. However, the inequality in accessible data volumes in today's attention economy is even greater.

Nevertheless, if properly used, blockchain technology is a possible means to reach the next level of human, social, cultural evolution. It can provide society with awareness and collective memory, if the slowness and significant energy consumption of today's blockchains can be overcome. It could be used to boost creativity, innovation, coordination, sustainability and resilience, hence, enable an entirely new, efficient and trustable organization of the world's societies at large.

Human evolution depends on the ability to coordinate people with diverse interests and goals. When genetic favouritism (giving advantages to relatives) was partially replaced by direct reciprocity ("an eye for an eye, a tooth for a tooth"), societies reached the next level of cooperation. In the past centuries, cultural evolution has further progressed with the implementation of more sophisticated cooperation mechanisms such as "indirect reciprocity" (i.e. reputation- and trust-based systems). Now, with the invention of the blockchain and similar technologies, the next level of society appears to be within reach, as it is possible to establish trust in a peer-to-peer way even between selfish actors, without the need of intermediary institutions.

---

[2]Overall, as I explain in my talk accessible under the URL https://www.youtube.com/watch?v=u-TCsFNnj54, the list of issues to be considered when using Blockchain technologies, includes: energy efficiency, over-regulation, too much determinism, over-commercialization, questions of governance and data ownership, crowding out of intrinsic motivation, danger of a post privacy world, large inequality, and a one-dimensional, utilitarian approach rather than a multi-dimensional one.

Blockchain is giving societies an unalterable ledger of our dealings with each other—a veritable registry, on the basis of which reputations can be assessed, and deceit can be unmasked. It is now possible to create collective awareness of how events are actually playing out and how they come about. Blockchain allows one to build a digital society, in which the legitimacy of interactions can be checked and verified.

Delft University of Technology has years of experience with primitive ledgers to record interactions. For instance, the Barter ledger[3] records who shared Internet bandwidth with whom. Even when interactions are anonymous, such as in Bittorrent peer-to-peer file sharing environments, using interaction records it is easy to identify and discourage unfair, non-reciprocal use of resources in the system (here: bandwidth).

So, what does this ultimately imply for the way we may all interact in future societies? With the concept of decentralized autonomous organizations (DAOs), blockchain technology can not only cure all sorts of blown-up bureaucratic structures, by coordinating people, resources, and processes in more transparent and efficient ways. It will even allow one to build a new form of socio-ecological, liberal, efficient and democratic kind of capitalism. This will consider externalities of everyone's activities on their environment and others by combining blockchain technology with the Internet of Things, creating a socio-ecological finance system.[4] In such a way, it is possible to boost a sustainable circular and sharing economy, with a variety of incentives, i.e. new socio-economic feedbacks.

Evolutionary biology shows that human language has evolved to give us the ability to talk about each other. This has boosted survival in a life-threatening world. Next, social intelligence evolved. Now, blockchain technology may create a new basis of truth and trust. A tamper-proof escape from lying, cheating, and hurting others would be a major leap forward in human evolution. We can do this now.

Today's world lacks memory and awareness of the reality we influence and which influences us. With blockchain technology this can now be changed. However, given that there are different ways of building a blockchain-based society, we must avoid to fall into the trap of a totalitarian post-privacy world, in which people might be restricted—and unnecessarily restrained in unfolding their knowledge, ideas, and talents. If we want to see a world with a level playing field for everyone, we need to insist on *responsible* blockchain innovations and on using distributed ledger mechanisms for the greater good, rather than allowing them to be usurped and harnessed by a very limited group of people for private interests.

It is important to figure out (e.g. by means of multi-player online games or Virtual Reality experiments) what information should be disclosed to whom and at what point in time, while avoiding harmful information asymmetries. Human dignity,

---

[3]https://repository.tudelft.nl/islandora/object/uuid:59723e98-ae48-4fac-b258-2df99d11012c?collection=education

[4]https://www.theglobalist.com/financial-system-reform-economy-internet-of-things-capitalism/

socio-economic diversity, and the outcomes of social self-organization may significantly depend on this.

The digital society we have in mind would offer protection and fair opportunities to all, while fostering collective intelligence, based on the sharing of knowledge and ideas. Openness, interoperability, fair access, and participatory opportunities would allow everyone to stand on the shoulders of others, thereby boosting a thriving society without avoidable shortages. This new digital age would empower everyone to be better informed and more innovative. With a subsidiary form of organization, it would allow everybody to participate in the co-creation of the spheres of life we care about, while helping us to coordinate our creative forces. By considering externalities, this can now be done in a way that minimizes harm to the environment and others while maximizing beneficial effects. So, what are we waiting for? Let's build the blockchain age together!

Jeroen van den Hoven is full professor of Ethics and Technology at Delft University of Technology and editor in chief of Ethics and Information Technology.

Johan Pouwelse is an associate professor of Computer Science at Delft University of Technology. He is founder of the TU Delft Blockchain Lab, see https://www.tudelft.nl/delft-blockchain-lab/

Dirk Helbing is full professor of Computational Social Science at ETH Zürich.

Stefan Klauser is a political scientist and fintech expert at ETH Zurich.

Chapter 14
From War Rooms to Peace Rooms: A Proposal for the Pro-Social Use of Big Data Intelligence

Dirk Helbing and Peter Seele

Digital technology has reached a supremacy and momentum hardly comparable with previous inventions. Even earlier game changers in the history of man such as the industrialization appear less impactful compared to the digital revolution. Both, utopian and dystopian futures appear possible[1]. Therefore, rather than using big data and artificial intelligence in "war rooms", we propose to use them in what we call "peace rooms" to promote collaboration, participation, pro-social applications, and peace.

It gets easily forgotten that the most disruptive potential today lies in the political, economic, social and cultural inequalities between digital super powers and the rest of society. Cold war, hot war, clash of cultures, disinformation and fake news, the decline of deliberative democracies in post-fact societies, massive automation, and new strategic alliances challenge established post World War II constellations and drive our societies to the brink of recession, civil war, and societal collapse, at least in some countries.

This article by Dirk Helbing and **Peter Seele** is an expanded version of a text that has appeared in Nature under the URL https://www.nature.com/articles/549458c and in The Globalist under the URL https://www.theglobalist.com/technology-big-data-artificial-intelligence-future-peace-rooms/

[1]Schmidt E, Cohen J (2013) *The new digital age. Reshaping the future of people, nations and business*. Knopf, New York.

D. Helbing (✉)
ETH Zurich, Zürich, Switzerland

TU Delft, Delft, Netherlands

Complexity Science Hub, Vienna, Austria
e-mail: dhelbing@ethz.ch

P. Seele (✉)
Universita della Svizzera Italiana, Lugano, Switzerland
e-mail: peter.seele@usi.ch

It would be naïve to assume that international intelligence agencies would not have various drastic scenarios ready somewhere in their folders. From what we know as informed public, measures have been taken and strategies prepared to deal with various kinds of catastrophic scenarios, some of which relate to our serious lack of sustainability. These scenarios are usually studied and managed in so-called "war rooms". Today, such war rooms are run with big data, using artificial intelligence and cognitive environments. In these environments, military experts and secret service people interact with advanced technical equipment through gestures and speech and analyze massive amounts of data and large numbers of scenarios. This considers huge knowledge bases and bodies of literature. Such war rooms may also support exploratory, team and emergency cognition, as well as consultation, discovery, collaborative decision-making, and guidance.

However, these analyses are typically based on past and current data. So, if war was the solution to certain problems in the past, war will be suggested as a solution in the future, even though innovations might bring peace. War room kind-of technology is also increasingly being used for strategic decision-making in big business: While the military would probably try to maximize power, businesses would maximize profit, but a thriving society needs the consideration of many goals, i.e. "value pluralism".[2] This requires a new approach—one that we will call "peace room".

So far, war rooms have informed military decision makers how to best attack other armies or countries or how to defend them from attacks. However, they have been much less successful in helping to build peace or create large-scale societal cooperation. In his seminal book on preventing deadly conflict, David Hamburg identified cooperation, coordinated international efforts, democratic institutions and socio-economic development as key ingredients to prevent war and terrorism.[3] Building on these findings our proposal for an internationally mandated, participatory "peace room" intends to contribute to the agenda of advancing pro-social uses to advance peace, sustainability and resilience using big data, artificial and human intelligence.[4]

---

[2]J. Van den Hoven (2005). Design for values and values for design. Information Age 4. 4–7.

[3]Hamburg, D. (2002). *No More Killing Fields- Preventing Deadly Conflict*. Rowman and Littlefield, Lanham.

[4]Gijzen, H., (2013). Big Data for a Sustainable Future. Nature 502, 38.

Kshetri N (2014). The emerging role of big data in key development issues: opportunities, challenges, and concerns. Big Data Soc 1(2):1–20. https://doi.org/10.1177/2053951714564227

Seele, P. (2016). Envisioning the Digital Sustainability Panopticon: A Thought Experiment how Big Data may help advancing Sustainability in the Digital Age. Sustainability Science. 11(5), 845–854. https://doi.org/10.1007/s11625-016-0381-5

Seele, P.& Lock, I. (2017). The game-changing potential of digitalization for sustainability: Possibilities, Perils, and Pathways. Sustainability Science. 12(2), 183–185. https://doi.org/10.1007/s11625-017-0426-4

14.1 Transparency and Participatory Opportunities are Important

The peace room should have access to the latest technology and powerful data sources. Legitimized by a democratic mandate, resulting from a transparent, participatory dialogue, properly trained and accredited peace room staff would have access to Big Data that has been collected for the purpose of running cities, countries, and disaster response.

To prevent misuse, the peace room would be monitored. The level of access may be determined by qualification and reputation (in terms of demonstrated responsible use). In any case, however, personal or other sensitive data will not be revealed to the staff. Political and non-governmental organizations, potentially also citizens[5] might propose topics or tasks (which we will call "missions"). In a next step, democratic voting could be used to prioritize missions. The compatibility of the missions with the mandate of the peace room would be checked by the executive and ethical boards, and regular reports would ensure transparency to the general public.

To be clear: a peace room is not about developing armed peace missions. It would rather serve to address global public issues threatening peace and prosperity from a global and doomsday-prevention perspective. The peace room is about the transition from war to peace and thus about reconciliation.[6] As compared to a war room setting, a "peace room" is characterized by a number of additional features such as: a higher degree of transparency (to reduce possible flaws and increase trust), a democratic framework of operation (for legitimacy), the use by interdisciplinary teams meeting international scientific standards (to achieve the integration of the best knowledge available), the supervision by ethical experts (to ensure responsible use and innovation), a multi-stakeholder and multi-perspective approach (to find solutions that work for everyone—as much as this is possible), and, in order to increase problem solving capacity, participatory opportunities for civil society (by means of NGOs, citizen science, and/or crowd sourcing).

For example, a peace room could be connected with participatory platforms such as UN OCHA, Global Pulse, the Deliberatorium,[7] and/or Nervousnet, which could connect with citizens and provide services to them. Inspired by the FuturICT initiative, Nervousnet aims at generating crowd-sourced data, using smartphones and other Internet of things sensors run by citizens.[8]

---

[5]See, for example, the opinion poll in the Netherlands exploring what scientists should work on: http://www.wetenschapsagenda.nl/national-science-agenda/how-the-national-science-agenda-will-be-developed/?lang=en

[6]Pennisi, E. (2012). From War to Peace. Science 336, Issue 6083, pp. 841. https://doi.org/10.1126/science.336.6083.841

[7]Klein, M. (2012). How to Harvest Collective Wisdom on Complex Problems: An Introduction to the MIT Deliberatorium. MIT Center for Collective Intelligence Working Paper No. 2012-04.

[8]Helbing, D. (2015). *The Automation of Society is Next: How to Survive the Digital Revolution.* CreateSpace Independent Publishing.

A peace room's scope of applications may span from transportation and logistics to smart cities and smart ports, to health, security and education. In the following, let us illustrate how a peace room—possibly in connection with other information platforms—could help to make our planet more resilient and sustainable quickly (besides other measures such as "democratic capitalism"[9] and "digital democracy"[10]).

14.2   Participatory Resilience and Sustainability

First of all, one could organize City Olympics, where cities around the world would regularly compete for the best environmental-friendly, energy-efficient, resource-saving and crisis-proof solutions. There would be different fields of competition, as well as various "weight classes" (small, medium-sized, and large cities, for example). This competition would involve science and engineering (R&D), but also business, politics, and the media.

Of course, peace rooms could support such activities, by providing data access and AI-based tools. Friendly competitions like these would also be important to mobilize the people to use resources more efficiently and to buy more environmentally-friendly products and technologies. Information about the best technologies, organizational principles and mobilization strategies could then be exchanged between the cities every other year. This strategy implies a combination of competition and cooperation between cities. Moreover, if the resulting innovations would be under the Creative Commons license and open source, the solutions could be easily further developed by everyone. This would lead to new businesses and a fast and widespread adoption of the best solutions.

Second, systemic resilience could be strengthened by diverse and decentralized/modular solutions as well as by bringing different local resources together efficiently with next generation sharing economy platforms. This became clear during an Earthquake resilience hackathon in San Francisco.[11] Participatory disaster response strategies could mobilize the full response capacity of society by enabling people to help themselves and help each other. One smartphone app ("Amigocloud") was proposed to report damaged infrastructures and other problems by uploading it to a map with peoples' smartphones in a geolocated way. Another app ("Helping hands") was offering the possibility to request help in the neighbourhood, offer support or resources, and coordinate supply and demand locally. Autonomous charging stations using solar panels ("Charge Beacons") would allow people to recharge their smartphones, which would—taken altogether—become important survival tools. Complementary, one could also consider to add energy-autonomous fab labs or maker spaces, in order to be able to locally fabricate objects and tools as needed.

---

[9]http://futurict.blogspot.de/2017/06/propositions-on-perspective-global.html

[10]Helbing, D. & Pournaras, E. (2015). Build Digital Democracy. Nature 527, 33–34. https://doi.org/10.1038/527033

[11]https://www.youtube.com/watch?v=OIlC7BQ84fE

Third, the financial and monetary system, which has been in crisis for years, should be replaced with a socio-ecological finance system ("finance system 4.0+"). This would work as follows: Using the sensors of the Internet of Things,[12] which are also in our smartphones, one could measure the impact of our actions on the environment and other people, for example, with a participatory measurement platform such as Nervousnet. This would enable one to quantify "negative externalities" such as noise, CO_2, and all sorts of waste. Similarly, "positive externalities" such as cooperation, education, health, and the recycling of resources could also be measured. Using blockchain technology—similar to the one behind the digital cryptocurrency "Bitcoin"—social value could then be generated in a decentralized way.[13] For example, various externalities could be assigned a price or value. In this way, a multi-dimensional incentive or financial system would be created, which would be suited for the real-time control of complex systems. With suitable incentives, the financial system could then be aligned with our constitutional and cultural values and with environmental requirements. In this way, externalities would be internalized in an innovative way, and new market forces would be unleashed, which would boost the development of a much more sustainable circular and sharing economy. This could provide a high quality of life for more people with fewer resources. It would also benefit companies, citizens and society alike.

All of the above proposals are perfectly compatible with democracy and the fundamental values of our society. Such an approach would not only come with increased legitimacy, but could also benefit the innovation capacity of diverse stakeholders and the general population, e.g. support NGOs and grassroots engagement besides politics and business. We believe that broader deliberation and participation as well as value pluralism will generally lead to improved solutions and greater public benefits. This is also very much compatible with the open discourse and deliberative democracy that Habermas demanded, leading to a balance of interests of different groups of society—an important condition for the possibility to unfold everyone's potential. Given the many pressing problems of our world, which have the potential to seriously affect billions of lives on Earth, it is now time to change the paradigm of how strategic decisions—particularly in crisis situations—are made, leading us from "war rooms" to "peace rooms".

Dirk Helbing is Professor of Computational Social Science at ETH Zurich, affiliate professor at TU Delft, and an external faculty member of the Complexity Science Hub Vienna. He is an elected member of the German Academy of Sciences "Leopoldina" and serves in various committees addressing the opportunities and challenges of the digital transformation.

Peter Seele is Professor of Business Ethics at USI Lugano, Università della Svizzera italiana, where he directs the Ethics and Communication Law Center (ECLC).

---

[12]Peacock S (2014) How web tracking changes user agency in the age of big data: the used user. Big Data Soc 1(2):1–11. https://doi.org/10.1177/2053951714564228

[13]Dierksmeier, C. & Seele, P. J Bus Ethics (2016). https://doi.org/10.1007/s10551-016-3298-0

Chapter 15
New Security Approaches for the Twenty-First Century: How to Support Crowd Security and Responsibility

Dirk Helbing

How can we protect companies and people from violence and exploitation? How can we open up information systems for everyone without promoting an explosion of malicious activities such as cyber-crime? And how can we support the compliance with rule sets on which self-regulating systems are built?

These challenges are addressed by Social Information Technology based on the concept of crowd security. A self-regulating system of moderators and the use of reputation systems are part of the concept. Today's reputation systems, however, are not good enough. It is essential to allow for multiple quality criteria and diverse recommendations, which are user-controlled. This leads to the concept of "social filtering" as a basis of a self-regulating information ecosystem, which promotes diversity and innovation.

Better awareness can help to keep us from engaging in detrimental, unfair or unsustainable interactions. However, we also need mechanisms and tools to protect us from violence, destruction and exploitation. Therefore, can we build Social Information Technologies for protection? And how would they look like? The aim of such Social Information Technologies would be to avoid such negative interactions, organize (collective) support or get fairly compensated. Of course, we also need to address here the issues of cyber-security and of the world's peace-keeping approach. Let us start here with the latter.

This chapter by Dirk Helbing was first published on February 7, 2015, as FuturICT Blog under the URL http://futurict.blogspot.com.eg/2015/02/new-security-approaches-for-21st.html

D. Helbing (✉)
ETH Zurich, Zürich, Switzerland

TU Delft, Delft, Netherlands

Complexity Science Hub, Vienna, Austria
e-mail: dhelbing@ethz.ch

© Springer International Publishing AG, part of Springer Nature 2019
D. Helbing (ed.), *Towards Digital Enlightenment*,
https://doi.org/10.1007/978-3-319-90869-4_15

15.1   The "Balance of Threat" Can be Unstable

Like many, I have was raised in a period of cold war. Military threats were serious and real, but the third world war did not happen. This is generally considered to be a success of the "Balance of Threat" (or "Balance of Terror"): if one side were to attack the other, there would still be time to launch enough intercontinental nuclear warheads to eradicate the attacker. Given the "nuclear overkill" and assuming that no side would be crazy enough to risk elimination, nobody would start such a war.

However, what if this calculus is fundamentally flawed? There were quite a number of instances within a 60 years period, where the world came dauntingly close to a third world war. The Cuban missile crisis is just the most well-known, but there were others that most of us did not hear about. (see http://en.wikipedia.org/wiki/World_War_IIIWorld War III and http://www.theguardian.com/world/2014/apr/29/nuclear-accident-near-misses-reportRisks of nuclear accidents is rising). Perhaps, we have survived the tragedy of nuclear deterrence by sheer chance?

The alarming misconception is that only shifts in relative power can destabilize a "Balance of Threat". This falsely assumes that balanced situations, called equilibria, are inherently stable, which is actually often not the case. To illustrate, consider the simple experiment of a circular vehicle flow (see http://www.youtube.com/watch?v=Suugn-p5C1Mvideo): although it is apparently not difficult to drive a car at constant speed together with other cars, the equilibrium traffic flow will break down sooner or later. If only the density on the traffic circle is higher than a certain value, a so-called "phantom traffic jam" will form without any particular reason—no accident, no obstacles, nothing. The lesson here is that dynamical systems starting in equilibrium can easily get out of control even if everyone has good information, the latest technology and best intentions.

What if this is similarly true for the balance of threat? What if, this equilibrium is unstable?

Then, it could suddenly and unexpectedly break down. I would contend that a global-scale war may start for two fundamentally different reasons. Consider a simple analogue from physics in which a metal plate is pushed from two opposite sides. In the first situation, if either of the two sides holding the plate becomes stronger than the other, the metal plate will move. Hence, the spheres of influence will shift. The second possibility is that both sides are pushing equally strong, but they are pushing so much that the metal plate suddenly bends and eventually breaks.

Often when an international conflict emerges, an action from one side triggers a counter-action from the opposing side. One sanction is met by something else and vice versa. In this escalating chain of events, everyone is pushing harder and harder without any chance for either side to gain the upper hand. In physics example, the metal plate may bend or break. In practical terms, the nerves of a political leader or army general, for example, may not be infinitely strong. Furthermore, not all events are under their control. Thus, under enormous pressure, things might keep escalating and may suddenly get out of control, even if nobody wants this to happen, if everyone just wants to save face. And this is still the most optimistic scenario, one in which all actors act rationally, for which there is no guarantee, however.

In recent years evidence has accumulated to demonstrate that in human history many wars have occurred due to either of the instabilities discussed above. The FuturICT blog on the http://futurict.blogspot.ie/2014/10/complexity-time-bomb-when-systems-get.htmlComplexity Time Bomb described how war can result without aggressive intentions on either side. Furthermore, recent books have revealed that World War I resulted from an eventual loss of control—the outcome of a long chain of events—a domino effect that probably resulted from the second kind of instability. Moreover, conflict in the Middle East has lasted for many decades, and it taught us one thing: Winning every battle does not necessarily win a war (quoted in the movie "The Gatekeepers" by a former secret service chief). Similar lessons had to be learned from the wars in Afghanistan and Iraq. Therefore, a new kind of thinking about security is needed.

15.2 Limits of the Sanctioning Approach

Whilst sanctioning might in some cases create social order, it can also cause instability and escalation in others. In the conflict in the Middle East, punishment is unsuccessful—the punishee does not accept the punishment, because values and culture are different. In such cases, the punishment is considered to be an undue assault and aggression, and therefore a strong enough punishee will strike back to maintain his/her own values and culture. In this manner, a cycle of escalation ensues, where both sides further drive the escalation, each fuelled by their conviction they are doing the right thing. In such a situation, deterrence is clearly not an effective solution. In other words, it is not useful to organize security alliances among countries which share the same values, as this creates precisely the cultural blocks that are unable to exercise acceptable sanctioning measures and will therefore run into escalating conflicts that can result in wars. Instead we need a new, symmetrical security architecture that is suited for a multi-polar world able to deal with cultural diversity. What we need are new strategies and a new kind of thinking. We also need a suitable approach in face of newly emerging cyber-threats.

15.3 How to Manage a Multi-polar World?

In the past, we have had a world with a few superpowers and blocs of countries forming alliances with them. Whenever one of these countries would be under attack, they would be under the protection of the others belonging to the same bloc. After World War II, the United States of America and Russia were the only superpowers remaining. With the breakdown of the Warsaw pact, there remained just one superpower. China is now the strongest economic power in the world and with Russia's comeback to world politics through the conflicts in Syria and the Ukraine, we are now living in a multi-polar world. Such a world is not well controllable anymore, as the "Three-Body Problem" suggests. This problem originally refers to the interaction of three celestial bodies, for which chaotic dynamics may result despite the simple conservation laws of mechanics. So, how much more unpredictable would a multi-polar world be?

It becomes increasingly obvious that today no power (political or business) in the world is strong enough to play the role of a world police, and that we need a new security architecture. If this would be an architecture for the entire world, it would need to have a number of features: The classical security alliances (power blocks) would have to be overcome. In view of globalization, thinking from the perspective of nation states seems to make decreasing sense. Furthermore, the concept of a "Balance of Threat" would have to be replaced by a "Network of Trust." The concept would have to be symmetric and not based on exclusive rights or veto power. It would have to be based on a set of shared values, and whoever violates them would feel the joint response of all the other countries in the world, independently of who their classical alliances were. For this approach to work well, mutual trust would have to grow, which would require more transparency and less secrecy.

15.4   In the Emerging Digital Society, How Much Secrecy Is Still Essential?

I cannot give a definitive answer to this, but I do believe that secrecy in the right time, place and context may have some benefits (e.g. privacy). But how much opacity should public institutions acting behalf of their citizens be allowed to have? And for what time period? Will the concept of secrecy be feasible at all in the future? Certainly Wikileaks and the Snowden revelations raise the question of whether secrets can still be kept in a data-rich world. Moreover, secret services have often been accused of engaging in unlawful behaviour, which they claim is necessary, to get an inside view of the closed circles of terrorism and organized crime. However, it has been stressed by some that such a strategy may actually promote terrorism and crime, and undermine the legitimacy of secret services, or even the states or powers they are serving. Finally, the effectiveness of secret services has often been questioned, and also whether they do more good than harm.

15.5   What Alternatives Might We Have to Create a New Security Architecture?

In this context, it is relevant to consider that more than 95% of the knowledge of secret services derives from public sources. As ever more activities in the world now leave a digital shadow and become traceable in real time, couldn't the largest part of public security be produced by public services rather than secret services? This does not necessarily mean to close down secret services, but to open up more information for wider circles. For example, why shouldn't specially qualified and authorized teams at public universities develop the algorithms and do the data mining to identify suspicious activities? Thanks to their higher transparency, they are exposed to scientific criticism and public scrutiny and would therefore be able to deliver higher-quality results. Given the many mistakes one can make when mining data, this would probably reduce the risk of wrong conclusions and other undesirable side effects. I am convinced that a step towards more transparency could largely increase the perceived legitimacy of the security apparatus and also the trust of people in the activities of their governments and states.

Perhaps, some readers of this book will find the above proposal to build public security on public efforts absurd, but it's not. In many countries, the police have already started to involve the citizens in their search for criminals such as through public webpages displaying pictures of suspects, as well as using text messages and social media. "Crowd security" is just the next logical step. In fact, we might put this into a bigger picture. As we know, the Internet started off with ARPANET, a military communication network. Opening it up for civilian use eventually enabled the creation of the World Wide Web, which then triggered off entirely new kinds of business and the digital economy. With the invention and ubiquity of Social Media, a large proportion of us has become part of a world-spanning network. The volume and dynamics of the related digital economy has become so extensive that the military and secret services can often not keep up with it anymore and, hence, they are increasingly buying themselves into civilian business solutions. This clearly shows that a future concept to protect our society and its citizens must largely build on the power of the civic society.

15.6 Crowd Security Rather Than Super Powers

Let me give an example of a system, in which crowd security is surprisingly effective and efficient, and where it creates "civic resilience". In the late 1990s, I spent some time as a visiting scientist at Tel Aviv University with Isaac Goldhirsch. At that time I read in the tourist guide that the average age of people in the country was 32, so I thought "this trip may be my last, given all the suicide bomb attacks." But it turned out that the young average age was a result of the comparatively high birth rate in this country, and I found myself enjoying my stay in the Middle East immensely. Despite the daily threats, people seemed to have a positive attitude towards life.

One of the things that impressed me much was the way security at public beaches was achieved, all based on unwritten rules. Everyone knew that any bag at the beach might contain a bomb that could kill you. Bags with nobody around were considered to be particularly suspicious. But at a beach, there are always some people swimming, so unminded baggage is normal. In this situation, people solve the problem by forming an invisible security network. Upon joining the beach, everyone becomes part of this informal network and implicitly takes responsibility for what is going on. That is, everyone scans the neighbourhood for suspicious activities. Who has newly arrived at the beach? What kind of people are they? How do they behave? Do they know others? Where do they go, when leaving their baggage alone etc.? In this way, it is almost impossible to leave a bag containing a bomb without arousing the suspicions of other people. To the best of my knowledge, there were relatively few bomb explosions at the beaches.

I would like to term the above distributed security activity as "crowd security". We have recently learned about the benefits of "crowd intelligence," "crowd sourcing," and "crowd funding," so why not "crowd security"? In fact, the way societies establish and maintain social norms is very much based on a "peer punishment" of those who violate these norms. From raising eyebrows to criticizing others, or showing solidarity with someone who is being attacked, there is a lot one can do to support a fair coexistence of people. I recall that, during one of our summer schools on Lipari Island in Italy, one of our US speakers noted: "In my country, you cannot even distribute some flyers in a private mall without security stepping in, but nevertheless, there are shootings all the time. I am surprised that everything is so peaceful in the public space on this island: young people next to old ones, Italians next to all sorts of foreigners, and I have not even seen a single policeman all these days." Again, people seem to be able to sort things out in a constructive way.

15.7 How Then Can We Generalize This Within an International Context?

I have sometimes wondered if having less power might work better than having more. When having little power, you must be sensitive to what happens in your environment, and this will help you to adapt (thereby allowing self-regulation to work). However, if you have a lot of power, you wouldn't make a sufficient effort to find a solution that satisfies as many people as possible. You would rather prioritize your own interests and force everybody else to adapt. But this would not create a system-optimal solution. As the example of http://futurict.blogspot.ie/2014/10/guided-self-organizationmaking.htmlcake-cutting suggests, the outcome wouldn't be fair, and therefore not sustainable on the long run. Why this? Because if you were too powerful, you would not get honest answers anymore, and sooner or later you would make really big mistakes that take a long time to recover from. For good reasons, Switzerland does not have a leader. The role of the presidency is taken for a short time period and rotates. This is interesting, as it requires everyone to find a sustainable balance of interests that is supported by many and, hence, has higher legitimacy. But there are more arguments than this for a decentralized, bottom-up "crowd security" approach.

15.8 The Immune System as Prime Example

One of the most astonishing complex systems in the world is our immune system. Even though we are bombarded every day by thousands of viruses, bacteria, and other harmful agents, our immune system is pretty good in protecting us for usually 5–10 decades. This is probably more effective than any other protection system we know. And there is another even more surprising fact: in contrast to our central nervous system, the immune system is "decentrally organized". It is a well known fact that decentralized systems tend to be more resilient. In particular, while targeted attacks or point failures can shut down a centralized system, a decentralized system will usually survive the impact of attacks and recover. This is one reason for the robustness of the Internet—and also the success of Guerrilla defence strategies (whether we like this or not).

15.9 Turning Enemies into Friends

There is actually a further surprise: a major part of our healthy immune response is based on our digestive tract, which contains up to a million billions of bacteria—10 times more than our body has cells. These bacteria are not only important to make the contents of our food accessible to our body, while they split them up into ingredients to find food for themselves. The rich zoo of about a thousand different bacteria in us even forms an entire ecosystem, which is fighting dangerous intruding bacteria that do not match the needs of our body. Bacteria that were once our enemies have eventually been turned into our allies through a symbiotic relationship that has eventually emerged through an evolutionary process. My friend and colleague, Dirk Brockmann recently pointed out to me to the really amazing level of cooperation, which is the basis of all developed life and now studied in the field of hologenomics. In fact, humans as well came up with tricky mechanisms encouraging cooperation. These are often based on exchange, such as trade, and a system of mutual incentive mechanisms, which promote coordination and cooperation. Social Information Technologies are intended to support this.

So why don't we build our societal protection system and the future Internet in a way that is inspired by our biological immune system?

It appears that societies as well have something like a basic immune system. The peer-to-peer sanctioning of deviations from social norms is one example for this, which I already mentioned before. We now witness internet vigilantes or lynch mobs on the web, criticizing things that people find improper or distasteful. I acknowledge that lynch mobbing can be problematic and may violate human rights; this will require us to find a suitable framework. It seems that we are seeing here the early stage of the evolution of a new, social immune system. Rather than censoring or turning off social media as in some countries, we should develop them further to make them compatible with our laws and cultural values. Then systems like these could provide useful feedback that would help our societies and economy to provide better conditions, products and services.

The question is how do we best obtain a high level of security in a self-regulating economy and society?

In perspective, we might create a security system that is partly based on automated routines and partly on crowd intelligence. If I can illustrate this again with the example of the Internet: let's assume that servers which are part of the Internet architecture, would autonomously analyze the data traffic for suspicious properties, but—in contrast to what we are seeing today—we would not run centralized data collection and data analytics. (Our brain certainly does not record and evaluate everything that happens in our immune system, including the digestive tract, but our body is nevertheless protected pretty well.) In case of detected suspicious activities, a number of responses are conceivable, for example: (1) the execution of the activity could be put on hold, while the sender is asked for feedback, (2) the event could trigger an alert to the sender or receiver of the data, a local administrator, or to a public forum, whatever seems appropriate. The published information could be screened by a crowd-based approach, to determine possible risks (particularly systemic risks) and to take proper action. While actions of type (1) would be performed automatically by computers, algorithms, or bots, actions of type (2) would correspond to the complementary crowd security approach. In fact, there would be several levels of self-regulation by the crowd, as I describe later. One may also imagine a closer meshing of computational and human-based procedures, which would mutually enhance each other.

15.10 Managing the Chat Room

We have seen that information exchange and communication on the web has quickly evolved. In the beginning, there was no regulation or self-regulation in place at all. These were the times of the Wild Wild Web, and people often did not respect human dignity or the rights of companies. But police and other executive authorities were also experimenting with new and controversial Internet-based instruments, such as Internet pillories to publicly name people.

All in all, however, one can see a gradual development of improved mechanisms and instruments. For example, public comments in news forums were initially published without moderation, but this spread a lot of low-quality content. Then, comments were increasingly assessed for their lawfulness (e.g. for respecting human dignity) before they went on the web. Then, it became possible to comment on comments. Now, comments are rated by the readers, and good ones get pushed to the top. The next logical step would be to rate commentators and raters. We can see thus the evolution of a self-regulatory system that channels the free expression of speech into increasingly constructive paths. I believe it is possible to reach a responsible use of the Internet based on principles of self-regulation. Eventually, most malicious behaviour will be managed by automated and crowd-based mechanisms such as the reporting of inappropriate content and reputation-based placements. A small fraction will have to be taken care of by a moderator, such as a chat room master and there will be a hierarchy of complaint instances to handle the remaining, complicated cases. I expect that, in the end, only a few cases will remain to be decided at the court, while most activities will be self-governed by social feedback loops in terms of sanctions and rewards by peers.

The above mechanisms will also feed-back from the virtual to the real world, and we will see an evolution of our over-regulated, inefficient, expensive and slow legal system into one that is largely self-regulating, more effective and more efficient. Here we may learn from the way interactive multi-player online games or Interactive Virtual Worlds are managed, particularly those populated by children. One of my colleagues, Seth Frey, has pointed me to one such example, the Penguin Club. To keep bad influences away from children, communication and actions within the Penguin Club world are monitored by administrators. As the entire population of Penguin Club users is too large to be mastered by a single person, there are several communities run on several servers, i.e. the Penguin Club world is distributed. Moreover, as every administrator manages his or her community autonomously, these may be viewed as parallel virtual worlds. This provides us with an exceptional opportunity to compare different ways of governance. Our study is far from being completed, so I just want to mention this much: It turns out that, if vandalism is automatically sanctioned by a robotic computer program, this tends to suppress creativity and results a boring world. This is reminiscent of the many failed past attempts to create well-functioning, liveable cities managed in a top-down way.

Returning to the virtual world of Penguin Club, I certainly don't want to argue in favour of vandalism, but I want to point out the following: the most creative and innovative ideas are, by their very nature incompatible with established rules, and it requires human judgement to determine, whether they should be accepted or sanctioned. This has an interesting implication: we may actually allow for different rules to be implemented in different communities, as they may find different things to be acceptable or not. This will eventually lead to diverse Interactive Virtual Worlds, which gives people an opportunity to personally choose their fitting world(s).

15.11   Embedding in Our Current Institutional System

Of course, we need to make sure to stay within the limits of the constitution and fundamental laws, such as human rights and respect for human dignity. Such decision may require difficult moral judgements and require particular qualifications of the "judge," the administrator of the gaming community or chat room. So it does make sense to have a hierarchy of such "judges" based on their qualification to decide difficult matters in an acceptable and respected way. These arbiters would be called "community moderators".

How would a "hierarchy of competence" emerge among such community moderators? This would be based on previous merits, i.e. on qualifications, contributions, and performance. Decisions would be rated both from the lower and the upper level. Over sufficiently many decisions, this would determine who will be promoted—always for a limited amount of time—and who will not. If the punished individual accepts the sentence of the arbiter, the moderation procedure is finished, and the sentence is published. Otherwise, the procedure continues on the next higher level, which is supposed to spend more effort on finding a judgement compatible with previous traditions, to reach a reasonable level of continuity and predictability.

Whoever asks for a judgement process (or revision) would have to come up for the costs (depending on the system, this might also be virtual money, such as credit points). Judgements on higher levels would become more expensive, and for the sake of fairness, fees and fines will not correspond to a certain absolute amount of money, but to a certain percentage of the earnings made in the past, for example, in the last 3 years. For example, in Switzerland, such a percentage-based system is successfully applied to traffic fines.

Only when the above-described self-regulation fails to resolve a conflict of interest over all judgement instances of the Interactive Virtual World would today's central authorities need to step in. One might even think that many of today's legal cases could be handled in the above crowd-based way of conflict resolution, and that today's judges would then only form the highest hierarchy. This would fit the system of self-regulation proposed above into our current organization of society. I expect the resulting procedures to be effective and efficient. The long duration of many court cases could be dramatically cut down. In other words, new community-based institutions of self-regulation should be able to help resolve the large majority of conflicts of interest better than existing institutions. I see the role of courts, police, and military mainly to help restore a balance of interests and power, when other means have failed. In this connection, it is important to remember that control attempts in complex systems often fail and tend to damage the functionality of the system rather than fixing it in a sustainable way. Therefore, I don't think that these institutions should try to control what happens in society.

15.12   Ending Over-Regulation

I believe that over time the principles of self-regulation will replace today's over-regulated system. A hundred years ago, only a handful of laws were made in the United Kingdom in one year. Now, a new regulation is put into practice every few hours. In this way, we have arrived in a system with literally tens of thousands of regulations. Even though we are supposed to, nobody can know all of them (but ignorance does not excuse us). Moreover, many laws are often revised shortly after their first implementation.

Even lawyers don't know all laws and regulations by heart. If you ask them, whether one thing is right or the opposite, they will usually answer: "it depends." So, we are confronted with a system of partially inconsistent over-regulation, which puts most people into a situation, where they effectively violate laws several times a year—and they even don't know in advance how a court would judge the situation. This creates an awkward arbitrary element in our legal system. While some people get prosecuted, others get away, and this creates an unfair system, not just because some can afford to have better lawyers than others.

However, this is not the only way an unfair situation is created, while our law system intends just the opposite i.e. to ensure a system that doesn't generate advantages for some individuals, companies, or groups. So what is the problem? Whenever a new law or regulation is applied, it requires some people or companies to adapt a lot, while others have to adapt just a little. This creates advantages for some and disadvantages for others. Powerful stakeholders would make sure a new law will fit their needs, such that they must adapt only a little, while their competitors would have to adapt much. Hence, the new law will make them again more powerful. However, even if we had no lobbying to reach law-making tailored to particular interest groups, the outcome would be similar. Just the stakeholders who profit most would vary more over time. The reason is simple: If N regulations are made and p is the probability that you have to adapt little, while $(1 - p)$ is the chance that you have to adapt a lot, the probability that you are a beneficiary k times is pk $(1 - p)(N - k)$. In other words, there is automatically a very small percentage of stakeholders who benefits from regulations enormously, while the great majority is considerably disadvantaged relative to them. Putting it differently: The homogenization of the socio-economic world comes along with a serious problem: the more rules we apply to everyone, the fewer people will find this world not well fit to their needs. And this explains a lot of the frustration among citizens and companies, not just in the European Union.

Only a highly diverse system with many niches governed by their own sets of rules allows everyone to thrive. Interestingly, this is exactly how nature works. It is the existence of numerous niches that allows many species to survive, and new ones to come up. For similar reasons, socio-economic diversity is an important precondition for innovation, which is important for economic prosperity and social well-being. Nature is much less governed by rules than today's service societies. For example, recent discoveries of "epigenetics" revealed that not even the genetic code is always read in the same way, but that its transcription largely depends on the biological and social environment.

Thus, how to build socio-economic niches, in which people can self-organize according to their own rules, within the boundaries of our constitution? Can we find mechanisms that promote social order, but allow different communities to co-exist, each one governed by their own sets of values and quality criteria? Yes, I believe, this is possible. Social Information Technologies will help people and companies to master the increasing levels of diversity in a mutually beneficial way (see Chap. 17 in this book and my book "The Automation of Society Is Next: How to Survive the Digital Revolution"). Furthermore, reputation systems can promote cooperation. If they are multi-dimensional, pluralistic, and community-driven, they can offer a powerful framework for social self-regulation, which provides enough space for diversity and opportunities for everyone.

15.13   Pluralistic, Community-Driven Reputation Systems

Here I want to elaborate a bit more on another important component of the "social immune system", namely reputation systems. These days, reputation and recommender systems are spreading over the Web which stresses their value and function. People can rate products, news, and comments, and they do! If they make the effort, there must be a reason for it. In fact, Amazon, Ebay, Tripadvisor and many other platforms offer other recommendations in exchange. Such recommendations are beneficial not only for users, who tend to get a better service, but also for companies, since a higher reputation allows them to sell a product or service at a higher price. However, it is not good enough to leave it to a company to decide, what recommendations we get and how we see the world. This would promote manipulation and undermine the "wisdom of the crowd" leading to bad outcomes. It is, therefore, important that recommender systems do not reduce socio-diversity. In other words, we should be able to look at the world from our own perspective, based on our own values and quality criteria. Only then, when these different perspectives come together, can collective intelligence emerge.

As a consequence, reputation systems would have to become much more user-controlled and pluralistic. Therefore, when users post ratings or comments on products, companies, news, pieces of information, and information sources (including people), it should be possible to assess not just the overall quality, but also different quality dimensions such as the physical, chemical, biological, environmental, economic, technological, and social qualities. Such dimensions may include popularity, durability, sustainability, social factors, or how controversial something is. It is, then, possible to identify communities based on shared tastes (and social relationships).

We know that people care about different things. Some may love slapstick comedies, while others detest them. So, it's important to consider the respective relevant reference group, and this might even change depending on the respective role we take, e.g. at work, at home, or in a circle of friends. To take this into account, each person should be able to have diverse profiles, which we may call "personas". For example, book recommendations would have to be different, if we look for a book for ourselves, for our family members, of for our friends.

15.14   Creating a Trend to the Better

Overall, the challenge of creating a universal, pluralistic reputation system may be imagined as having to transfer the principles, on which social order in a village is based, to the global village, i.e. to conditions of a globalized world. The underlying success principle is a merit-based matching of people making similar efforts. This can prevent the erosion of cooperation based on "indirect reciprocity," as scientists would say. For this approach to play out well, there are a number of things to consider: (1) the reputation system must be resistant to manipulation attempts; (2) people should not be terrorized by rumours; (3) to allow for more individual exploration and innovation than in a village, one would like to have the advantages of the greater freedoms of city life—this requires sufficient options for anonymity (to an extent that cannot challenge systemic stability).

First, to respect the right of informational self-determination, a person would be able to decide what kind of personal information (social, economic, health, intimate, or other kind of information) it makes accessible for what purpose, for what period of time, and to what circle (such as everyone, non-profit organizations, commercial companies, friends, family members, or just particular individuals). These settings would, then, allow selected others to access and decrypt selected personal information. Of course, one might also decide not to reveal any personal information at all. However, I expect that having a reputation for something will be better for most people than having none, if it would help find people who have similar preferences and tastes.

Second, people should be able to post their comments or ratings either in an anonymous, pseudonymous, or personally identifiable way. But pseudonymous posts would have, for example, a 10 times higher weight than anonymous ones, and personal ones a 10 times higher weight than pseudonymous ones. Moreover, everyone who posts something would have to declare the category of information: is it a fact (potentially falsifiable and linked to evidence allowing to check it), an advertisement (if there is a personal benefit for posting it), or an opinion (any other information). Ratings would always have the category "opinion" or "advertisement". If people use the wrong category or post false information, as identified and reported by, say, 10 others, the weight of their ratings (their "influence") would be reduced by a factor of 10 (of course, these values may be adjusted). All other ratings of the same person or pseudonym would be reduced by a factor of 2. This mechanism ensures that manipulation or cheating does not pay off.

Third, users would be able to choose among many different reputation filters and recommender algorithms. Just imagine, we could set up the filters ourselves, share them with our friends and colleagues, modify them, and rate them. For example, we could have filters recommending us the latest news, the most controversial stories, the news that our friends are interested in, or a surprise filter. So, we could choose among a set of filters that we find most useful. Considering credibility and relevance, the filters would also put a stronger weight on information sources we trust (e.g. the opinions of friends or family members), and neglect information sources we do not want to rely on (e.g. anonymous ratings). For this, users would rate information sources as well, i.e. other raters. Then, spammers would quickly lose their reputation and, with this, their influence on recommendations made.

Users may not only use information filters (such as the ones generating personalized recommendations), but they will also be able to generate, share, and modify them. I would like to term this approach "social filtering." (A simple system of this case has been implemented in Virtual Journal).

Together, the system of personal information filters would establish an "information ecosystem," in which increasingly reliable filters will evolve by modification and selection, thereby steadily enhancing our ability to find meaningful information. Then, the pluralistic reputation values of companies and their products (e.g. insurance contracts or loan schemes) will give a quite differentiated picture, which can also help the companies to develop customized and more useful/successful products. Reputation systems are therefore advantageous for both, customers and producers. Customers will get better offers, and producers can take a higher price for better quality, leading to mutual benefit.

15.15 Summary

Social Information Technologies for protection might be imagined to work like a kind of immune system, i.e. a decentralized system that responds to changes in our environment and checks out the compatibility with our own values and interests. If negative externalities are to be expected (i.e. if the value of an interaction would be negative), a protective "immune response" would be triggered.

Part of this would be an alarm system, a kind of "radar" that alerts a user of impending dangers and makes him or her aware of them. In fact, the "Internet of Things" will make changes—both gains and losses—measurable, including psychological impacts such as stress, or social impacts, such as changes in reputation or power. Social Information Technologies for protection would help people to solidarize themselves against others who attack or exploit them. A similar protection mechanism may be set up for institutions, or even countries. Such social protection ("crowd security") might often be more efficient and effective than long-lasting and complicated lawsuits. Of course, protection by legal institutions would exist, but lawsuits would become more like a last resort than a first resort, for when social protection fails, e.g. when there is a need to protect someone from organized crime. Note that already a suitably designed reputation system would be expected to be quite efficient in discouraging certain kinds of exploitation or aggression, as it would discourage others from interacting with such people or companies, which would decrease the further success of those who trouble others.

Chapter 16
Homo Socialis: The Road Ahead

Dirk Helbing

When I started the publication project with Herbert Gintis on the "homo socialis" (Gintis and Helbing 2015), *the most important motive for me was to trigger a scientific debate. So, from my perspective, our joint paper on the "homo socialis" is not to be seen as an end point or eternal truth, but as the starting point of a new theory for socio-economic systems. In this comment, I will expand on my paper with Herbert Gintis, and I will use the opportunity to present some further thoughts and materials.*

16.1 Evolution of the "Homo Socialis"

Since my PhD days, I have wondered how it was possible that psychology, sociology and economics were all claiming to model the decision-making of people, while at the same time using (at least partly) different sets of models and assumptions. So, overcoming the divide between economics and the other social sciences seemed necessary (Eckel and Sell 2015), and this has been part of my research agenda ever since. My collaboration project with Herbert Gintis was born out of a project with Thomas Grund and Christian Waloszek that was part of this agenda, where we put the "homo economicus" to the test. Thomas, Christian and I simulated the

---

This chapter by Dirk Helbing was first published as an open access article under the same title in the Review of Behavioral Economics 2 (1–2), 239–253 (2015) and is accessible via the URL https://www.nowpublishers.com/article/Details/RBE-0032

D. Helbing (✉)
ETH Zurich, Zürich, Switzerland

TU Delft, Delft, Netherlands

Complexity Science Hub, Vienna, Austria
e-mail: dhelbing@ethz.ch

evolutionary dynamics that is sometimes claimed to be the reason for the existence of the "homo economicus." Our computer simulation model distinguished utilities from payoffs and made four assumptions, none of which directly implied other-regarding behavior (Grund et al. 2013; Helbing 2013a):

- Agents decide according to a best-response rule that strictly maximizes their utility function, given the behaviors of their interaction partners (their neighbors).
- The utility function considers not only the own payoff, but gives a certain weight to the payoff of the interaction partner(s). The weight is called the "friendliness" and set to zero for everyone at the beginning of the simulation.
- Friendliness is a trait that is inherited (either genetically or by education) to offspring. The likelihood to have an offspring increases exclusively with the own payoff, not the utility function. The payoff is assumed to be zero, when a friendly agent is exploited by all neighbors (i.e. if they all defect). Therefore, such agents will have no offspring.
- The inherited friendliness value tends to be that of the parent. There is also a certain mutation rate, but it does not promote significant levels of friendliness.

What did our computer simulations of the biological evolution of utility maximizing agents tell us? For many parameter combinations, the outcome was indeed a "homo economicus," as most economists would expect. Surprisingly, however, there was also an area of the parameter space, where a "homo socialis" with other-regarding preferences emerged, namely when offspring grew up next to their parents (see Fig. 16.1). Given that most humans actually do raise their children at home, this is quite intriguing. It is also interesting that, while in the beginning of our agent-based computer simulations other-regarding preferences are disadvantageous, they can achieve higher payoffs after several dozen generations.

Fig. 16.1 Outcome of an evolutionary simulation of human preferences (from Grund et al. 2013)

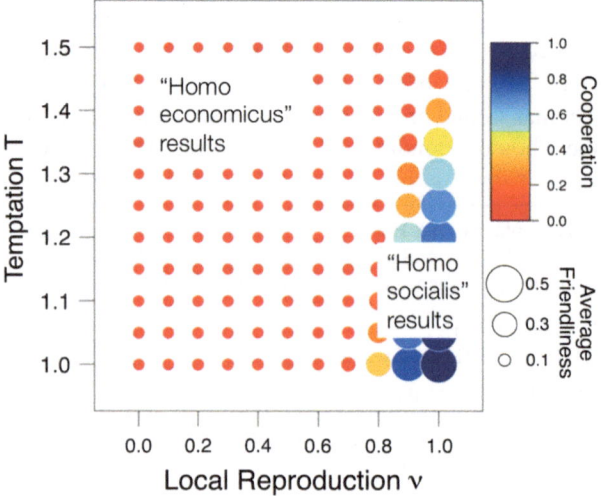

When offspring are raised close to their parents, we find not only other-regarding behavior (cooperation), but also the emergence of a "homo socialis" with other-regarding preferences. This provides a theory explaining experimental findings on fairness preferences, conditionally cooperative behavior, and individual utility functions (Fischbacher et al. 2001). The results of the computer simulation further prove that the consideration of "externalities" (i.e., of external effects of decisions and actions) can yield a better system performance and benefit everyone, which hints towards superior organization principles for economies, as they now become possible by the Internet of Things with emergent sensor networks that will make it possible to measure externalities of all kinds.[1]

Remarkably, with the "homo socialis," there exists a second reference point besides the "homo economicus" that an analytical economic theory could be built around. So, is it possible that economic theory was developed around the wrong reference point? Indeed, many human interactions are indicative of the "homo socialis" rather than the "homo economicus." Moreover, could it be that the difference between the behavior of the "homo economicus" and the "homo socialis" is so big that it cannot anymore be treated as a small deviation from the "homo economicus"—an approximation error, which averages out over sufficiently many decisions? If so, the average behavior would effectively deviate from mainstream economic theory, and we would need a new economic theory, and new economic institutions as well (see, for example, Helbing 2013a—in particular the discussion of the social preference literature relating to economic laboratory experiments).

In fact, in an ever more networked world, where consumers interact and buy products through social media, the concept of separate decision-making is decreasingly plausible. If decisions become more interdependent than they used to be, theory must increasingly account for the implications of "networked minds," and representative agent theory will have to be replaced by theories of complex dynamical systems (Helbing and Kirman 2013). As Gallegati (2015) points out: "sociality implies interaction, which produces externalities." And as Lewis (2015) underlines, Hayek noticed already early on that economic systems should be studied as complex systems. This would have to include explanations of emergent collective phenomena and novel system properties resulting from individual-level interactions as mentioned by Lewis (2015) and Nowak et al. (2015). The question is: are these just gradual improvements or will the implications of complex social interdependencies be as exciting as the discovery of quantum mechanics or of the theory of relativity? Is economic theory perhaps at a turning point?

One might, of course, argue that rational choice theory has already been adapted long ago to account for individual preferences. This is reflected by individual utility functions. However, it is too simple to say that economics is the study of choice under constraints with given preferences, and to leave it to sociologists to explain the individual preferences. As some of the comments to our paper have rightly pointed

---

[1] see http://papers.ssrn.com/sol3/papers.cfm?abstract_id=2583391 and http://futurict.blogspot.ch/2014/09/creating-making-planetary-nervous.html

out (Hechter 2015; Hodgson 2015; Isaac 2015), without a theory how to theoretically determine these individual preferences, ideally in advance, rational choice theory is pretty incomplete and of limited use. But I believe that a theory describing how individual preferences and utility functions come about can actually be formulated. In fact, in the study of Grund et al. (2013), individual utility functions are an outcome of an evolutionary process, and they are a result of interactions in the past.

Having said this, let me respond to some of the comments to the paper of Herbert Gintis and myself, as I lay out further ideas on the evolution of human decision-making and its—as I believe—rather interesting implications. The replies to our paper contain many thoughtful comments, and I agree with many of them. They have highlighted different aspects that certainly deserve attention in the further debate about a core analytical theory for the social sciences. The great majority of these points actually turn out to have played important roles in my email exchange with Herbert Gintis when we worked on our common paper. Apparently not all of these points made it into our paper, but this gives me an excellent opportunity to present them here.

16.2   Limitations of Equilibrium Theory

The question to what extent economic systems can be assumed to be in equilibrium, has been in the center of scientific debates (Ormerod 2015; Witt 2015). I have been questioning equilibrium approaches myself (Helbing and Balietti 2010). In fact, they may not always be suitable to describe (decisions and learning in) quickly changing environments.

Therefore, my paper with Herbert Gintis certainly does not want to imply that equilibrium can always be assumed. It just likes to say that the analysis of stationary points can be insightful, and that the classical equilibrium concept can be extended in ways that consider social aspects. Generally, however, a system of equations of the kind $F_k (x_1, \ldots, x_i, \ldots x_n) = 0$ with a solution $(y_1, \ldots, y_i, \ldots, y_n)$ may just reflect the stationary state of a dynamical set of equations $dx_k/dt = F_k (x_1, \ldots, x_i, \ldots x_n)$. In such a case, it makes sense to determine the eigenvalues of the matrix with the elements dF_k/dx_i in the stationary point $(y_1, \ldots, y_i, \ldots y_n)$. If all of these eigenvalues are negative (or have negative real values), deviations from the stationary point $(y_1, \ldots, y_i, \ldots y_n)$ as they may be caused by perturbations of the system, would tend to decrease over time. Consequently, the system would be driven towards the stationary point $(y_1, \ldots, y_i, \ldots y_n)$—at least, if there is just one stationary solution, or if the perturbation is sufficiently small. In this case, the system will be usually well described by its equilibrium $(y_1, \ldots, y_i, \ldots y_n)$.

However, if at least one of the eigenvalues is positive (or has a positive real value), the system will eventually be driven away from the stationary solution $(y_1, \ldots, y_i, \ldots y_n)$, and it might end up in a different stationary solution. In systems of non-linear dynamical equations, non-stationary behaviors such as oscillatory or chaotic solutions can be possible as well, which is well-known from complexity theory (Haken 2012; Nowak et al. 2015). Moreover, even if (the real values of) all eigenvalues are negative (i.e. all variables tend to follow a damped dynamics) such that the system is expected to behave stable, it might happen that new perturbations occur before previous ones have disappeared. A nice example for this effect, which is sometimes called "convective instability," is the "bullwhip effect" that is sometimes observed in supply chains (Helbing and Lämmer 2005). A similar effect might be relevant for financial markets (where it may create bubbles or crashes) and for other socio-economic systems experiencing high innovation rates.

In fact, non-equilibrium behaviors of socio-economic systems are common. A socio-economic system in equilibrium cannot produce the innovations needed to adapt well to a changing world. It is the nature of many innovations that they destabilize a previously established equilibrium and promote a new structure, process or system behavior. Innovations tend to increase diversity, and diversity tends to accelerate innovation (Helbing et al. 2005b). Moreover, a diverse economy is related with a high gross national product (Hidalgo et al. 2007; Page 2008). Innovations are, therefore, desirable. The related process of differentiation is an important non-equilibrium feature of successful economies, and heterogeneity should therefore be a key ingredient of economic models (Gallegati 2015).

So far, however, it is still a theoretical challenge to understand the conditions that create particular kinds of inventions, and also the conditions supporting their spreading. A model that allows one to grasp innovation as system-immanent process, considering effects of randomness, would be highly desirable. However, this is difficult because innovations may be disruptive in the sense that they do not just improve the performance of a previously existing technology or procedure, but also create entirely new quality dimensions or functionalities. As one of the comments put it, strategy spaces cannot be specified ahead of time (Wolpert 2015). Innovation is open-ended (Lewis 2015). It can transcend the existing socio-economic system and may not be captured by a closed system of equations. Therefore, certain aspects related to novelty-generation and emergence, such as "radical" (Ormerod 2015; Lewis 2015) or "fundamental" uncertainty (Helbing 2013b)—where the probabilities and/or utilities of certain events cannot be enumerated anymore—are difficult to account for.

Nevertheless, evolutionary models (Helbing 1992; Young 1993; Weibull 1997; Helbing et al. 2005b; Gintis 2009) considering mutations are trying to grasp at least some of the process of novelty-generation. But certain outcomes can only be understood by co-evolutionary processes, for which correlations are essential. Then, the common factorization assumption used to derive the mean-value equations underlying many representative agent models cannot be applied (an application would eliminate relevant emergent phenomena). The entire concept of "correlated equilibria" (or "resonant correlations", as Vernon Smith 2015, likes to call them) would obviously not work, if correlations were not relevant.

16.3 Role of Randomness

I fully agree that randomness may have significant effects (Smith 2015). For example, it may lead to the emergence of cooperation between strangers (Grund et al. 2013). The emergence of the "homo socialis" that I mentioned earlier would not occur without "errors" or "noise" (Smith 2015). In fact, the transition from the "homo economicus" to the "homo socialis" needs a coincidence of random mutations of several behaviors in a certain neighborhood. Initially, such mutations are dysfunctional and do not pay off, i.e. they turn out to be "mistakes." But beyond a certain critical group size, friendliness pays off.

Another example for the relevance of noise has been given in a recent experimental paper (Mäs and Helbing 2017). There, we have shown that a deterministic micro-level theory—the myopic best response rule—describes 96% of all individual decisions correctly, but it surprisingly fails to reproduce the outcome of the collective dynamics. This can happen when small deviations matter, i.e. when the stationary (or "equilibrium") solution is unstable. Then, tiny perturbations can sometimes trigger dramatic amplifications through cascade effects, which may even have system-wide impacts (Helbing 2013b). Note that heterogeneity in a system may have similar implications as well (Gallegati and Kirman 1999). In such cases, local interaction effects and correlations can be so relevant that they sometimes produce very different outcomes from what a representative agent model predicts (Gallegati 2015).

Interestingly, adding noise to decision models can increase their predictive power. For example, in contrast to the above-mentioned deterministic best response model, a stochastic version corresponding to the multi-nomial logit model (McFadden 1973), reproduces the distribution of macro-level experimental outcomes much better (Mäs and Helbing 2017).

16.4 An Alternative Foundation of Decision Theory

In many cases, it is possible to model the role of noise by stochastic games (Wolpert 2015) or by following a master equation approach (Weidlich 2000). The latter can also be used as an alternative starting point of choice theory (Helbing 1995). This line of thought to substantiate utility theory can be summarized as follows (Helbing 2004): Let us assume choice options $x1, x2, \ldots, xi, \ldots, xn$, and choice probabilities $p(xi)$ (which may change over time). Then we can define a transformation via $p(xi) = N*\exp(ß*vi)$, where N is a normalization factor and ß is a noise parameter. This transformation with the exponential function may be justified by the logarithmic law of psycho-physics, underlying our senses or the geometric averaging that people tend to perform. There is also a relationship to the gross-canonical distribution in physics (Helbing 1995). If the parameter ß were infinity, this would correspond to a deterministic choice of the option with the highest utility, but in realistic settings, ß is finite.

The values vi, which I will call utilities, can be ordered to define a preference scale, which reflects different choice probabilities. An interesting implication is the following: Let us assume a lottery choosing x1 with probability q and x2 with probability $(1 - q)$. The expected utility of this new choice option x3 would be $v3 = q * v1 + (1 - q) * v2$. The choice probability would then be proportional to exp $(ß^* v3) = \exp \{ß^* [q^* v1 + (1 - q)^* v2]\} = \exp (ß^* q^* v1)^* \exp [ß^* (1 - q)^* v2] = p(x1)q^* p(x2)(1 - q)$ with $p(x1) = \exp (ß^* v1)$ and $p(x2) = \exp (ß^* v2)$. This is the well-established and widely used Cobb-Douglas function.

In many cases, one needs, of course, to consider joint probabilities $p(xi,xj) = p(xi|xj)p(xj)$. Then, the Bayesian formula follows directly from probability theory. We can also transform the conditional probabilities $p(xi|xj)$ of choosing xi given xj—without limitation of generality we may write $p(xi|xj) = N^* \exp(ß^* uij)$. Then, uij can be split up into an asymmetric and a symmetric part: $uij = sij + aij$ with $sij = (uij + uji)/2 = sji$ and $aij = (uij - uji)/2 = -aji$. One possible specification of the asymmetric part would be $aij = vi - vj$, where vi can again be called utility. Sij may be interpreted as similarity between two options xi and xj. $dij = \exp(-ß^* sij) = dji$ can be used to define distances.

Conditional probabilities are necessary to understand not only conditional choice (which is, for example, relevant to understand social norms), but also sequences of actions, which are part of many social roles, and they are relevant for correlated equilibria as well. Turn-taking and its evolution is a nice example for this (Helbing et al. 2005a).

The above foundation of utility-based decision theory has the appeal that it does not require to assume a computation or even the maximization of utility. It just assumes choice probabilities. When conditional probabilities are considered, one can also model dependencies on irrelevant alternatives and intransitive preferences scales (Isaac 2015; Ormerod 2015), such as different restaurant choices (Hodgson 2015). In fact, conditional preferences are important to understand the variability of preferences over time. For example, when we have eaten, we are not hungry anymore, and other things become more preferable. I will come back to this saturation-kind of time dependence of individual preferences below.

Another nice example is a competitive game on a circle, where one gets the highest payoff, if one is a step ahead of the others (Frey and Goldstone 2013). This produces a constant forward movement. For example, in business, one always likes to be a step ahead of the competition. This causes constant change. But after a few steps, one might end up again where one started. In fact, "fashion cycles" are a well-known phenomenon (Helbing 1995).

16.5 Beyond Rational Choice

In agreement with some of the comments (Goldstone 2015; Nowak et al. 2015), I am convinced that the above decision theory needs further extensions. There is a lot of evidence that evolution has equipped humans with different incentive and reward systems, for example, sexual pleasure (to ensure reproduction), possession-related satisfaction (to survive in times of crises), appreciation of novelty (to explore opportunities and risks), or empathy-related satisfaction—sympathetic fellow-feeling, as Vernon Smith (2015) calls it. These establish different motivational factors, which—I claim—should not be aggregated into a single utility function, but would be better represented by different dimensions of utility.

These different utilities cannot be perfectly traded against each other, and their relative importance may change quickly, thereby changing also our preference scales. In other words, human behavior results from different drivers, which dominate for some time and then give place to another. At each point in time, depending on the respective situational context, we prioritize a certain objective—here, the concept of "self-regulation" comes in (Lindenberg 2015). The switching between diverse objectives might be imagined to work similarly to the self-controlled traffic lights we have developed to serve vehicle queues at intersections (Lämmer and Helbing 2008). This self-control approach is based on the service of the most pressing local needs. Interestingly, when the externalities on neighboring intersections are taken into account, this distributed bottom-up control even outperforms classical attempts of top-down optimization (Helbing 2013a).

Let us discuss next how the apparently incompatible decision theories based on rational choice models and on the concept of decision heuristics (Gigerenzer et al. 2000; Gilovich et al. 2002) may be related to each other. It is plausible to me that people try to increase their different rewards (see the "hedonic goal" mentioned in Lindenberg 2015), and that they learn various heuristics for this, to improve their turnouts. It also makes sense to assume that a heuristic is selected depending on the situational context of a decision, such that framing matters (Lindenberg 2015). In contrast to the utility-maximizing approach of rational choice theory, heuristics do not necessarily result in optimal choices. However, they are time- and energy-efficient, and on average they work well, given sufficient opportunities to learn. Therefore, after a long enough learning time, the application of good heuristics would come pretty close to the maximization of a utility function. In other words, on an aggregate level, rational choice theory would be a good approximation of heuristic-based decision-making (but multiple utility dimensions for different, non-aligned reward systems and the switching between them would still have to be taken into account). In such a framework, rational decision-making may be seen as an emergent, approximate outcome, depending on the decision context (Gallegati 2015).

In fact, I am convinced that we can understand the diverse reward systems as results of (co-)evolutionary processes, and that the decision heuristics and their application can be explained as a result of reinforcement learning, given certain cognitive abilities. The ERC MOMENTUM project I am currently leading is trying to elaborate such an approach, based on agent-based computer simulations of cognitive agents with a virtual brain. These simulations distinguish processes on three different time scales: (i) decision-making, (ii) learning, and (iii) biological evolution. They involve genetic inheritance under mutations and reinforcement learning in an environment, where individuals compete for different kinds of rewards and individual success influences reproduction rates and the likelihood to be imitated. The ultimate ambition of the MOMENTUM project is to explain the emergence of reward systems, individual and collective intelligence, social behavior, and culture from first principles.

Furthermore, to understand collective intelligence, it is important to consider the social nature of individuals. "Networked minds" (Grund et al. 2013) allow for parallel information processing, knowledge sharing, etc. Then, not everyone has to evaluate all pieces of information relating to a certain problem (such as identifying the best insurance contract). It is enough if everyone evaluates some information and people then compare their conclusions with each other. (In fact, we don't read the details of all insurance contracts, before we choose one, but we ask some colleagues and friends we trust, and follow up some of their recommendations by further in-depth analysis. This is something not well represented by a theory of independent decision-making.) Putting it differently, collective intelligence allows individuals to process information in a distributed way, and to jointly find solutions that are better than each individual one. An important precondition for this is diversity, i.e. the fact that individuals often do not decide and behave in a representative way (Page 2008).

16.6 Gene–Culture Co-evolution

This brings us to the subject of gene–culture co-evolution, the understanding of which requires concepts such as cultural and multi-level selection (Lewis 2015). Determining to what extent individual preferences result from genetic inheritance as compared to cultural transmission by learning will certainly require further scientific studies. Universal facial expressions (Ekman and Friesen 1971) probably support the genetic inheritance of certain cultural abilities, but many other aspects such as religious values and beliefs may be just transmitted culturally. Imitation (Helbing 1992, 1995), teaching, and learning play a similarly important role for inheriting culture, as genetic inheritance plays it for the spreading of physiological capabilities. There must be a reason why most human offspring stay with their parents for almost two decades. This actually suggests a high relevance of cultural transmission.

However, when trying to understand human behavior, the role of biology can certainly not be ignored. Evolution determines our physiological capabilities. Our brain determines our cognitive ones. Cognitive abilities influence our behavior, our social institutions, and our reproduction, hence, evolution as well. In other words, we probably have a co-evolution of physiological, cognitive and social abilities. In fact, in certain cases it is not so clear whether a behavior is genetically inherited or

culturally spread. For example, is a preference for fairness and cooperation genetically or culturally inherited, or both? The capacity to speak, evolving together with
language use, is an interesting example for a co-evolution between physiological and
cultural abilities. I also expect that the cognitive capacity for empathy (being able to
put oneself into the shoes of others, see Lindenberg 2015) is genetically transmitted,
while education determines how we use it.

16.7 Importance and Origin of Morality

The above considerations are also relevant for another important subject, which has
been highlighted by one of the comments, namely learning to self-restrain (Smith
2015). It has been rightly pointed out that "formal legal rules are insufficient to
generate emergent coordinated actions; informal moral rules of promise-keeping and
truth-telling are needed as well." (Lewis 2015; see also Hodgson 2015). Yes, moral
judgments are not simply expressions of an individual's interests, preferences,
sentiments or beliefs, but a matter of doing the right things, even if one doesn't
like them. And I agree that the ability to consider moral rules is part of what makes us
human. In particular, I concur with the statements that "norms are a glue of
societies," as Michael Hechter (2015) points out, and that the "moral legitimacy of
the legal system in the eyes of citizens is crucial" (Hodgson 2015; see also
Lindenberg 2015).

The theory of correlated equilibria offers a partial understanding of some of these
issues. For example, it allows one to explain the emergence of social conventions
(Helbing 1992; Young 1993), social norms (Helbing and Johansson 2010), or turn-
taking without a "choreographer" (Helbing et al. 2005a; Goldstone 2015). Generally, social conventions and social norms help to improve coordination and to reduce
transaction failures (Winter et al. 2012). They may change the conditional choice
probabilities or even the choice set (when norms are "internalized"). However, the
concept of correlated equilibria is certainly not giving a full picture. On the one hand,
contents of moral values are hard to capture by means of theories. On the other hand,
norms are often stabilized ("cemented") by institutions such as police and jurisdiction, religion and culture.

But can we at least understand the origin of morality by quantitative models? This
is in fact the case. A partial answer is given by one of our agent-based computer
simulations (Helbing et al. 2010). It studies a social dilemma situation, in which
people can choose between four different strategies: (1) cooperate and punish
defectors, (2) just cooperate, while avoiding costly punishment, (3) defect, or
(4) defect while punishing other defectors. One may call type (1) "moral" and type
(4) "immoral" or "hypocritical" behavior. The simulation outcomes for this setting,
when assuming the imitation of better-performing behaviors of interaction partners,
is quite interesting: When everyone interacts with everybody else or with randomly
chosen interaction partners, corresponding to a representative agent model, a

"tragedy of the commons" results, where most individuals defect, while defection is not punished. In contrast, in a spatial setting where everyone interacts with the direct neighbors, moral behavior can emerge, i.e. a widespread cooperation with a punishment of defectors. (Therefore, both the first- and second-order free-rider dilemmas are solved.) This is due to homophily: "birds of a feather", i.e. similar strategies, cluster together. As a consequence, moralists don't have to compete with cooperators, but interact with defectors, such that costly punishment can succeed and spread. Local interactions and the co-evolution of punishment and cooperation are key to success. But the evolution of morality has, of course, further facets: it also involves deliberation, which requires higher-level intellectual abilities; and it also concerns the evolution of particular cultures and social institutions.

16.8 Role of Data and Experiments

Finally, I agree with the comment that we need better data even more than better theories (Macy 2015). Therefore, the role of computational social science (Lazer et al. 2009) deserves to be stressed a lot more. I believe quick scientific progress of socio-economic theories will crucially depend on the establishment of a circular feedback between theory and empirical or experimental evidence (Eckel and Sell 2015): data allow one to validate and calibrate or even empirically derive socio-economic models, but theories can also help one to identify interesting decision experiments (Helbing and Yu 2010) and to set up better measurement processes.

Besides lab and web experiments, Big Data about human activities will play a much bigger role in future socio-economic research (Conte et al. 2011). This ranges from the behavior of financial markets (Preis et al. 2012) over mobility patterns (Song et al. 2010) or daily activities (Golder and Macy 2011) to the spreading of culture (Schich et al. 2014). I also agree that we need to pay more attention to the socio-economic interaction networks (Schweitzer et al. 2009; Hechter 2015; Macy 2015; Nowak et al. 2015), as they can have dramatic influence on the system behavior (Helbing et al. 2010). The activities of my research team are, in fact, trying to bring these aspects together. For this, we have developed the Open Data Search Engine "Living Archive" (http://livingarchive.inn.ac, https://github.com/bitmorse/livingarchive), the NodeGame platform for Web experiments (http://www.nodegame.org/preview/, https://github.com/nodeGame), and the Virtual Journal platform to identify relevant scientific literature across disciplinary boundaries (http://vijo.inn.ac, https://github.com/bitmorese/vijo). These activities subscribe to an open source spirit enabling a community-based effort. The aim of the FuturICT initiative (http://www.futurict.eu) is to develop this on a global scale. We can do this together and thereby create a collective knowledge base that cuts across disciplinary boundaries. It would be great, if we could even establish a collective (problem-solving) intelligence, which goes beyond an additive approach.

Will you be part of this?

Acknowledgment The author acknowledges support by the ERC Advanced Investigator Grant 'Momentum' (Grant No. 324247).

References

Conte, R., Gilbert, N., Bonelli, G., Helbing, D.: FuturICT and social sciences: Big data, big thinking. Z. Soziol. **40**(5), 412–413 (2011). http://www.soms.ethz.ch/research/pub/ZfS_futureICT_social_sciences

Eckel, C., Sell, J.: Searching for "Homo Socialis:" A comment on Gintis and Helbing. Rev. Behav. Econ. **2**, 61–66 (2015)

Ekman, P., Friesen, W.V.: Constants across cultures in the face and emotion. J. Pers. Soc. Psychol. **17**, 124–129 (1971)

Fischbacher, U., Gächter, S., Fehr, E.: Are people conditionally cooperative? Evidence from a public goods experiment. Econ. Lett. **71**(3), 397–404 (2001)

Frey, S., Goldstone, R.L.: Cyclic game dynamics driven by iterated reasoning. PLoS ONE **8**(2), e56416 (2013)

Gallegati, M.: From the *homo economicus* to the *homo socialis*. Rev. Behav. Econ. **2**, 67–76 (2015)

Gallegati, M., Kirman, A.: Beyond the Representative Agent. Elgar, Cheltenham (1999)

Gigerenzer, G., Todd, P.M., The ABC Team: Simple Heuristics That Make Us Smart. Oxford University Press, Oxford (2000)

Gilovich, T., Griffin, D., Kahneman, D.: Heuristics and Biases: The Psychology of Intuitive Judgment. Cambridge University Press, Cambridge (2002)

Gintis, H.: Game Theory Evolving. Princeton University Press, Princeton, NJ (2009)

Gintis, H., Helbing, D.: Homo socialis: An analytical core for sociological theory. Rev. Behav. Econ. **2**, 1–59 (2015)

Golder, S.A., Macy, M.W.: Diurnal and seasonal mood vary with work, sleep and daylength across diverse cultures. Science. **333**, 1878–1881 (2011)

Goldstone, R.: Homo socialis and homo sapiens. Rev. Behav. Econ. **2**, 77–87 (2015)

Grund, T., Waloszek, C., Helbing, D.: How natural selection can create both self- and other-regarding preferences, and networked minds. Sci. Rep. **2**, 1480 (2013). http://www.nature.com/srep/2013/130319/srep01480/full/srep01480.html

Haken, H.: Synergetics. Springer, Berlin (2012)

Hechter, M.: Why economists should pay heed to sociology. Rev. Behav. Econ. **2**, 89–92 (2015)

Helbing, D.: A mathematical model for behavioral changes by pair interactions. In: Haag, G., Mueller, U., Troitzsch, K. G. (eds.) Economic Evolution and Demographic Change. Formal Models in Social Sciences, pp 330–348. Springer, Berlin, also see http://papers.ssrn.com/sol3/papers.cfm?abstract_id=2413177 for the initial formulation in 1990. http://www.soms.ethz.ch/research/pub/pairinteractions.pdf (1992)

Helbing, D.: Quantitative Sociodynamics. Stochastic Methods and Models of Social Interaction Processes. Kluwer Academics, Dordrecht (1995). 2nd edition: Springer, Berlin (2010). https://doi.org/10.1007/978-3-642-11546-2

Helbing, D.: Dynamic decision behavior and optimal guidance through information services: Models and experiments. In: Schreckenberg, M., Selten, R. (eds.) Human Behaviour and Traffic Networks, pp. 47–95. Springer, Berlin (2004). http://www.santafe.edu/research/working-papers/abstract/9268ffd21afbd598b22f9695a9e15c34/

Helbing, D.: Economics 2.0: The natural step towards a self-regulating, participatory market society. Evol. Inst. Econ. Rev. **10**(1), 3–41 (2013a). http://arxiv.org/abs/1305.4078

Helbing, D.: Globally networked risks and how to respond. Nature **497**, 51–59 (2013b). http://www.nature.com/nature/journal/v497/n7447/full/nature12047.html

Helbing, D., Balietti, S.: Fundamental and real-world challenges in economics. Culture **76**(9–10), 399–417 (2010). http://www.saha.ac.in/cmp/camcs/Sci_Cul_091010/17%20Dirk%20Helbing. pdf

Helbing, D., Johansson, A.: Cooperation, norms, and revolutions: A unified game-theoretical approach. PLoS ONE **5**(10), e12530 (2010). http://www.plosone.org/article/info%3Adoi% 2F10.1371%2Fjournal.pone.0012530

Helbing, D., Kirman, A.: Rethinking economics using complexity theory. Real-World Econ. Rev. **64**, 23–52 (2013). http://www.paecon.net/PAEReview/issue64/HelbingKirman64.pdf

Helbing, D., Lämmer, S.: Supply and production networks: From the bullwhip effect to business cycles. In: Armbruster, D., Mikhailov, A.S., Kaneko, K. (eds.) Networks of Interacting Machines: Production Organization in Complex Industrial Systems and Biological Cells, pp. 33–66. World Scientific, Singapore (2005). http://www.worldscibooks.com/engineering/ 5938.html

Helbing, D., Yu, W.: The future of social experimenting. Proc. Natl. Acad. Sci. USA (PNAS). **107** (12), 5265–5266 (2010). http://www.pnas.org/content/107/12/5265

Helbing, D., Schönhof, M., Stark, H.-U., Holyst, J.A.: How individuals learn to take turns: Emergence of alternating cooperation in a congestion game and the prisoner's dilemma. Adv. Complex Syst. **8**, 87–116 (2005a). http://www.worldscinet.com/acs/08/0801/ S0219525905000361.html

Helbing, D., Treiber, M., Saam, N.J.: Analytical investigation of innovation dynamics considering stochasticity in the evaluation of fitness. Phys. Rev. E. **71**, 067101 (2005b). http://scitation.aip. org/getabs/servlet/GetabsServlet?prog=normal&id=PLEEE8000071000006067101000001& idtype=cvips&gifs=yes

Helbing, D., Szolnoki, A., Perc, M., Szabó, G.: Evolutionary establishment of moral and double moral standards through spatial interactions. PLoS Comput. Biol. **6**(4), e1000758 (2010). http:// www.ploscompbiol.org/article/info:doi%2F10.1371%2Fjournal.pcbi.1000758

Hidalgo, C.A., Klinger, B., Barabasi, A.-L., Hausmann, R.: The product space conditions the development of nations. Science **317**, 482–487 (2007)

Hodgson, G.M.: A Trojan horse for sociology? Preferences versus evolution and morality. Rev. Behav. Econ. **2**, 93–112 (2015)

Isaac, A.G.: Comment on "*Homo Socialis*: An analytical core for sociological theory". Rev. Behav. Econ. **2**, 113–121 (2015)

Lämmer, S., Helbing, D.: Self-control of traffic lights and vehicle flows in urban road networks. JSTAT. P04019 (2008). http://www.iop.org/EJ/abstract/1742-5468/2008/04/P04019

Lazer, D., et al.: Life in the network: The coming age of computational social science. Science **323**, 721–723 (2009). http://www.ncbi.nlm.nih.gov/entrez/eutils/elink.fcgi?dbfrom=pubmed& retmode=ref&cmd=prlinks&id=19197046

Lewis, P.: An analytical core for sociology: A complex, Hayekian analysis. Rev. Behav. Econ. **2**, 123–146 (2015)

Lindenberg, S.: The third speed: Flexible activation and its link to self-regulation. Rev. Behav. Econ. **2**, 147–160 (2015)

Macy, M.W.: Big theory: A Trojan horse for economics? Rev. Behav. Econ. **2**, 161–166 (2015)

Mäs, M., Helbing, D.: Random deviations improve micro-macro predictions. An empirical test. Sociol. Methods Res. **1**, 31 (2017)

McFadden, D.: Conditional logit analysis of qualitative choice behavior. In: Zarembka, P. (ed.) Frontiers in Econometrics, pp. 105–142. Academic, New York (1973)

Nowak, A., Andersen, J., Borkowski, W.: Dynamics of socio-economic systems: Attractors, rationality, and meaning. Rev. Behav. Econ. **2**, 167–173 (2015)

Ormerod, P.: A comment on Gintis and Helbing "Homo Socialis: An analytical core for sociological theory". Rev. Behav. Econ. **2**, 175–182 (2015)

Page, S.E.: The Difference: How the Power of Diversity Creates Better Groups, Firms, Schools, and Societies. Princeton University Press, Princeton, NJ (2008)

Preis, T., Kenett, D.Y., Stanley, H.E., Helbing, D., Ben-Jacob, E.: Quantifying the behavior of stock correlations under market stress. Sci Rep. **2**, 752 (2012). http://www.nature.com/srep/2012/121018/srep00752/full/srep00752.html

Schich, M., Song, C., Ahn, Y.-Y., Mirsky, A., Martino, M., Barabasi, A.-L., Helbing, D.: A network framework of cultural history. Science **345**, 558–562 (2014)

Schweitzer, F., Fagiolo, G., Sornette, D., Vega-Redondo, F., Vespignani, A., White, D.R.: Economics networks: The new challenges. Science **325**, 422–425 (2009)

Smith, V.L.: Adam Smith: *Homo socialis*, yes; social preferences, no; reciprocity was to be explained. Rev. Behav. Econ. **2**, 183–193 (2015)

Song, C., Qu, Z., Blumm, N., Barabasi, L.A.: Limits of predictability in human mobility. Science **327**, 1018–1021 (2010)

Weibull, J.W.: Evolutionary Game Theory. MIT Press, Boston, MA (1997)

Weidlich, W.: Sociodynamics. CRC Press, Boca Raton (2000)

Winter, F., Rauhut, H., Helbing, D.: How norms can generate conflict: an experiment on the failure of cooperative micro-motives on the macro-level. Soc Forces. **90**(3), 919–946 (2012). http://sf.oxfordjournals.org/content/90/3/919.full?sid=57731ded-53ef-4ed2-92fd-1121f08b5427

Witt, U.: Sociology and the imperialism of economics. Rev. Behav. Econ. **2**, 195–202 (2015)

Wolpert, D.: The gaping holes in social science. Rev. Behav. Econ. **2**, 203–210 (2015)

Young, H.P.: The evolution of conventions. Econometrica **61**(1), 57–84 (1993)

Chapter 17
Social Mirror: More Success Through Awareness and Coordination

Dirk Helbing

In our daily lives, we may face three kinds of situations: win–win, win–lose, or lose–lose. Social Information Technologies can help us make the best out of these. They can filter and integrate information for us in real-time, and make it understandable for us, thereby turning raw data into useful information and actionable knowledge. A Social Mirror, for example, could improve our awareness of opportunities and risks related to our possible decisions and actions, while a Social Adapter could support our coordination with people having different interests and backgrounds, empowering us to interact and collaborate successfully.

As everyone knows, many problems result when people or companies don't care about the impact of their decisions on others. In fact, this may deteriorate the situation of everyone, as witnessed by so-called "tragedies of the commons," or mutually damaging conflicts. Polluted environments and run-down public infrastructures are well-known problems of this kind. How to overcome such problems? How to promote more responsible behaviors and sustainable systems?

The classical approach is to invent, implement and enforce new legal regulations. But people don't like to be ruled, they can't handle many laws, and they often find ways around them. As a consequence, laws are often not effective. Neither did free markets manage to overcome "tragedies of the commons" or destructive conflicts. Hence, we are faced with unintended problems such as overfishing and environmental degradation, but there are many more problems of this kind. Here, we propose a completely different approach: the Social Mirror. In the following, I will

This contribution by Dirk Helbing is related to the patent "Interaction Support Processor" accessible via the URL https://patentscope.wipo.int/search/en/detail.jsf?docId=WO2015118455

D. Helbing (✉)
ETH Zurich, Zürich, Switzerland

TU Delft, Delft, Netherlands

Complexity Science Hub, Vienna, Austria
e-mail: dhelbing@ethz.ch

© Springer International Publishing AG, part of Springer Nature 2019   201
D. Helbing (ed.), *Towards Digital Enlightenment*,
https://doi.org/10.1007/978-3-319-90869-4_17

explain, how Social Information Technologies can help us to grasp the complexity of our world and manage it better.

17.1   What's the Problem?

In complex systems such as our economy or society, interactions between the system components (such as individuals, computer systems, companies, etc.) can lead to unexpected, "emergent" outcomes. These outcomes can be favorable (such as self-organized cooperation or social order), but also unfavorable, as in the case of traffic jams, crowd disasters, financial crises, or "tragedies of the commons." Could we solve—or at least mitigate—these problems by new information and communication technologies and, if so, how? The answer is: if we had to come up for the external-ities caused by our decisions and actions, then individual and collective interests would become aligned. As a consequence, we shouldn't run so easily into situations where our behavior causes avoidable damage to others and the environment, and finally to us. Having to pay for negative externalities would also discourage inno-vations that would otherwise create large-scale damage to the benefit of a few. Just think of "toxic assets" as an example, i.e. innovative financial derivatives that almost made the financial system collapse. Unfortunately, many digital innovations are exploiting us as well. How to change this?

Let us assume there exists a "value function" that quantifies how desirable a certain "state" of the system is for the system's components. Here, system compo-nents can be anything from people over companies to robots or computer algorithms. The value function will usually depend on contextual variables such as the behaviors of other system components. It may also depend on previous states, i.e. the system's history.

In general, when two system components interact (where one may also be the environment), several outcomes are possible:

- *Case I (lose–lose situation):* The interaction wouldn't be favorable for any of the interacting components (i.e. it would lead to a less valued system state). In this situation, the interaction should be avoided. Segregation or decoupling strategies are suitable solutions. To prevent unfavorable interactions, *situational awareness* of the lose–lose situation is crucial. The *Social Mirror* described in this chapter may serve to achieve such awareness.

- *Case IIa (bad win–lose situation):* The interaction would be favorable for one component, but not for the other interacting component. If the benefit for one component is lower than the disadvantage for the other, the interaction should be avoided as in case I, since it would deteriorate the overall systemic outcome. However, as one component has a selfish interest in the interaction, measures are needed to *protect* the other component from exploitation. Here, awareness combined with a collective protection mechanism can help (see, for example, Chap. 15 in this book).
- *Case IIb (good win–lose situation):* The interaction would again be favorable for one component, but not for the other. However, if the benefit for one component is higher than the disadvantage for the other, the other component can be compensated for by a *value transfer* such that the interaction becomes profitable for both. This requires a *(Micro-) Payment System*. The transactions may be supported by a *Social Adapter,* as described in this chapter below. Again, one component must be protected from exploitation by the other as in case IIa.
- *Case III (win–win situation):* The interaction would be favorable for both components. In this case, the interaction should be performed. Nevertheless, a *value transfer* may still be carried out to reach a fair(er) sharing of profits by the interacting components. Again, a (Micro-) Payment System would serve this purpose.

While coordination or transaction failures can often be subsumed under case I, case II corresponds to conflicts of interest. In the following, I will describe *Social Information Technologies* to support favorable kinds of interactions and avoid unfavorable ones. In particular, I propose technological systems that could help us to determine the value ("success") of an interaction, to raise situational awareness for profitable or unfavorable interactions, to support a more successful execution of such interactions, and to support a value transfer, which increases individual and systemic benefits at the same time. In this way, one may significantly reduce the occurrence of systemic instabilities and cascade failures in complex techno-socio-economic-environmental systems. Before we discuss the *Social Adapter* at the end of this chapter, let us start with a description of the *Social Mirror*.

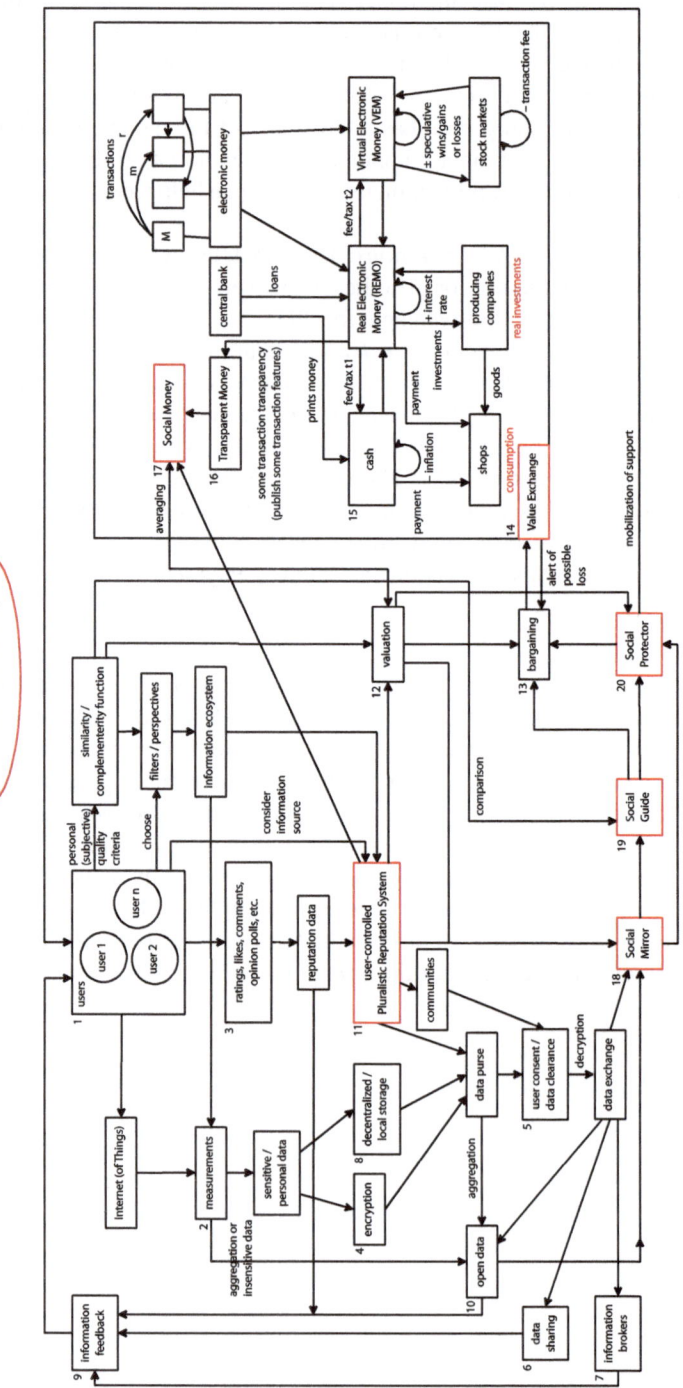

17.2 How the Idea Came About

One day, I was wondering, what one could do to turn the downward spiral leading to "tragedies of the commons" into an upward spiral creating a better world. Was it possible to find a trick that would promote more responsible and other-regarding behavior? How could ethical behavior survive and even triumph in a world of harsh competition?

I remembered a paragraph in a book I had read when I was young. In the World Trade Center, it said, there were always terrible queues in front of the elevators at the end of the working day. This annoyed people, until someone came up with a surprisingly simple solution: Putting mirrors in the waiting area would keep people entertained.

People want to be beautiful for others, I thought. They want to be recognized and loved. In fact, it is astonishing how much money people pay to be attractive, and to impress others. They spend thousands of dollars on expensive clothes, hairdressers, jewelry, watches, aesthetic surgeries, etc. And the mirror was the key invention that made them care so much about their appearance!

Was it perhaps possible to invent a mirror for "inner beauty"—a mirror showing the own behavior, it's impact on the environment and on others, and what people would think about the own behavior? Wouldn't this change people's behaviors to the better, as they would again want to impress others and increase their efforts to do things that others appreciate? In fact, people care a lot about their self-image. The problem is that we have no suitable tools to give us a good idea of our behavior and what it means to others. Making this visible and understandable could change everyone's behavior to the better—and could thereby change our world! A Social Mirror could give useful feedback to us. Such a device would make many of us want to behave nicer, to gain reputation and admiration.

So, what if everyone had a personalized avatar in the appearance of a beloved friend, guide, or genie (depending on your taste), who would give us hints and advice, what we might do in certain situations? An avatar that would help us to engage in more successful social interactions and economic exchange? This is exactly the idea behind the Social Mirror. Using information from the Internet and the sensor networks that the "Internet of Things" will provide, one could get a picture of one's own behavior and people's responses.

17.3 Mapping Human Culture

It has often been claimed that war is the mother of civilization. I don't buy this. War has been spreading the basis of civilizations, but with the Internet, this can now happen in different ways. However, it's a good question to ask what is actually the basis of civilizations? It's culture! And culture is largely a collection of established success principles, which may determine the evolution of a culture over thousands of years. Just take religious value systems, which have guided the decisions of believers since ages. It is therefore not exaggerated to say that cultures persist longer than steel. And it is more relevant for success than weapons. In other words, social forces can be stronger than physical force(s).

While we have found ways to understand physical forces, the social forces underlying cultural evolution are still not well understood, and that's why we often fail to master cultural diversity. Most cultural rules are nowhere written up, and we have often difficulties identifying and explaining them. But it does not have to stay like this. Just remember the cultural boost that occurred when people invented language and writing, or the phenomenal progress in biology, when the genetic alphabet was revealed. So, would something alike gene sequencing be possible for cultures?

What if we started to map the rules and success principles on which the cultures of the world are built, and developed a "cultural alphabet" describing our conventions, norms, values, and success principles? It would create a much better understanding of how our societies work. As a result, we could avoid a lot of affronts and stupid mistakes, where our behavior generates outcomes we did not want. In other words, with suitable digital assistance we could interact more successfully with people who are different from us. The question is, whether one could record culture in an automated way. Perhaps, this can be done in part, but I believe it would have to be complemented by a citizen science or crowd-sourced approach, as we know it from *Wikipedia, Zooniverse* or *OpenStreetMap*.

Google Glass is certainly a controversial device, as it records what is going on in the surrounding, but it could be also used for less privacy-invasive applications. Besides, "wearable computing" offers other, less intrusive ways of measuring social interactions. For sure it would be possible to extract relevant information about what pleases other people in a place and what annoys them. It would be possible, for instance, to study the mimic and gesture of people and do a sentiment analysis to assess a situation. Comparing this with tweets, blogs, and comments would allow one to consider local norms and cultural values. All of this could eventually go—in an anonymous and aggregated way—into *Wikidata* or some other *OpenCulture* database considering geolocations. In essence, such a "Culturepedia" platform would record the values and success principles of people in different places of the world. Since the way we appear to others depends on the context and location, it is important to compare (or contrast) one's behaviors with the respective local culture.

17.4 How Would the Social Mirror Work?

The overall approach of the Social Mirror is to enable more interesting and more fruitful interactions with the people and environments around us, and to avoid interacting with people, with whom it wouldn't be beneficial to interact with, i.e. to avoid lossful situations. While the Social Mirror would focus on a particular person, the Social Adapter discussed below would focus on the interactions with others.

If you like, the Social Mirror would tell you things like "Mr. X found today's conversation with you really pleasant," or "here you could have been a bit more diplomatic." Or "Mrs. Y was quite irritated when you said that she was very inspiring today." The Social Mirror could let you know that "it's fantastic that you have been doing more sports this week" or "it's great that you have consumed 5% less fuel this week" or "given your music taste, you might want to visit the concert of Z in a month just an hour from here."

If you like, the Social Mirror could also let you know that "you have just eaten c calories today, so you can still afford to have a desert." Or it could tell you that "if you go to work by car, it will take you 34 minutes at this time of the day, cost you x dollars, and produce y amount of carbon dioxide (CO_2). If you take public transport now, you will arrive in 37 minutes, but the amount of carbon dioxide will be one tenth of this, and it will cost you z dollars. Otherwise, if you go by bike, it will take you 1 hour and 3 minutes, burning y calories." After taking your decision, the Social Mirror would guide you where and when to catch the bus, change to the metro etc.

Of course, as the Social Mirror will eventually know your values and preferences well—it could also tell you: "Given the weather today, going to work by bike would suit you well". In other words, the Social Mirror would be tailored to you. You could, for example, tell it to give travel costs a weight w and travel time a weight $(1 - w)$, or it would just learn your preferences from your choices. Based on this, you could get personalized recommendations in an easily understandable and intuitive way.

In the following, we will assume that it will not be difficult to get or collect the data needed. In times of the "Internet of Things," there will eventually be an abundance of information about everything. Sensor networks may measure not only what can be perceived by our senses, but also radiation, chemicals, social links, psychological mood, and whatever may come to our mind. It would be concerning, however, to store a lot of personal data of many people in one place (or a few data centers). Therefore, the Social Mirror should be based on a decentralized information platform. In favor of informational self-determination, one should also make sure that we can determine the access to our personal data, e.g. by means of a *Personal Data Store*. Publicly available data would be aggregated or anonymized, and personal data disclosed only with our permission.

Considering this, we may also assume that the Social Mirror collects public opinions of people about subjects, objects, and information, and all sorts of information of interest, considering personal preferences, values and tastes. This includes ratings, reputational information, and contextual information of various kinds. Interactions with the environment would be important to consider as well.

The information could either be actively entered by individuals, or extracted automatically by text or sentiment analysis or other means. It would then be stored in a file associated with the respective subject or object, a specific, user-controlled data mailbox (the *Personal Data Store*). This information would be encrypted and, first of all, accessible to the person the information is collected about. Everyone could then decide with whom to share this information, and whose opinion one would like to listen to. It would also be possible to voice opinions anonymously. But one might also decide, not to listen to other opinions at all.

Furthermore, the Social Mirror would be able to give you selective feedback, e.g. only positive or only negative responses, or responses from certain kinds of people, whatever you choose. The Social Mirror could furthermore map out how environmentally or socially friendly our behavior is. Comparing this information with others might lead to a playful competition to care more about our environment and others. In fact, playfulness is an important design feature. We don't want someone to admonish us all the time what we are doing wrong. We want to have an enjoyable feedback that helps us to learn, overcome our weaknesses, and strengthen our strengths. As indicated before, such feedback could come from an avatar, who may appear like a friend, an action hero, a comic figure, or anybody we would like to listen to—and just at the times when feedback and advice is really appreciated.

17.5 Social Adapter and Intercultural Guide

Our current way of organizing social order is based on laws and norms and their formal or informal enforcement. This tries to make us all alike in a growing number of features, to standardize interactions and enable efficient exchange. While we appreciate efficiency and are often far from gentle to reach it, companies and people find it annoying to be restricted by laws and social norms. In fact, this undermines a lot of our creative potentials. It is no exaggeration to say that we are ruled by thousands of norms every day, and on top of this there are tens of thousands of applicable laws and regulations created every year (most of which we haven't even heard of). However, enforcing compliance with laws, rules, and regulations has serious side effects. It reduces socio-economic diversity, while diversity is key for innovation, societal resilience, and collective intelligence, and often for individual happiness, too. The problem is just that we have great difficulties to handle all the diversity around us in an efficient way.

However, consider that, every day, we happen to walk by hundreds or thousands of people we don't know. Some of them love the same kinds of books we like. Others share our taste of music or arts. Some may feel the way we do, and others might share our secret hopes, ideas and visions. Nevertheless, we don't talk to each other most of the time. Imagine all these missed opportunities! What, if we were able to notice and use them? We could all be so much happier, and so much more successful! Therefore, what is it that keeps us from exploring and using these opportunities? First of all, we don't often know who these people are, and second, conventions may discourage us to get in touch with them. Third, it often takes too much time to find out, who actually has what interests, talents and values—and how to avoid conflicts where people disagree.

The *Social Adapter* can be a solution to these problems. It will help us to interact more efficiently with people and companies having diverse interests, without assimilation and the need to reinforce social norms. The Social Adapter will rather make the rule sets pursued by others better understandable to us. It will support us in identifying matching interests and in bargaining value-increasing interactions to reach win–win situations. In other words, the Social Adapter will empower everyone to better cope with diversity, and to manage the balancing act between different kinds of interest.

The Social Adapter would be able to contrast individual opinions and behaviors with the opinions and behaviors in a particular location. In this way, it can give an idea what aspects of our own interests are fitting the local culture, and which ones not. The Social Adapter could also support us in finding individuals who have similar backgrounds and similar ways of thinking, thereby helping us to find friends and business partners.

For example, the *Social Adapter* would be able to tell you: "there is someone who admires your kind of work, would you like to talk to him?" Or: "there is someone, who has complementary interests and skill sets—would you like to get in touch with her?" Or "there is someone who has a problem with your points of views—beware of possible conflicts." Hence, the Social Adapter would be able to advise you on how to be more successful in certain situations. The Social Adapter could, therefore, be seen as a "personal guide".

In summary, the Social Mirror and Social Adapter could support the self-organization of individuals and companies in a decentralized and bottom–up way. Of course, the users should be able to determine themselves where, when and how to use the Social Mirror or Social Adapter in their decision-making. Sensitive information about individuals would be kept in their Personal Data Stores, and that information would be made accessible only to trusted companies or people that we want to share (some of) our data with. In fact, most transactions and data exchanges would not even need to reveal personal data at all. They could be processed in a distributed way by "trusted information brokers," i.e. on computer devices of independent third parties, who are no beneficiaries of the interaction and would process only small parts of the relevant data. In this way, a negotiation process could be performed even without the need to reveal sensitive personal or business data!

17.6   The Importance of Social Capital

Importantly, besides supporting personal actions and interactions, the Social Mirror and Social Adapter can also contribute to the creation of public goods, in particular social capital. Most of us have probably learned that money makes the world go 'round and all that matters is to have enough of it. Money is certainly a powerful invention, but there is more that contributes to economic, societal and cultural development. This includes human capital (like education), but also social capital. Thus, what is social capital? I define it as everything that results from social network interactions and can potentially be turned into a benefit. Examples are cooperativeness, public safety, a culture of punctuality, reputation, trust, respect, and power. While your own actions influence your social capital, you can't fully control it, which is in contrast to money. In many cases, you can't buy social capital (or only to a limited extent), but social capital creates value added. Interestingly, by doing certain things, you are not automatically entitled to get a certain amount of social capital. As we know from reputation and respect, these are things that are given to you by others. They depend on interaction effects.

Note that the amount of social capital also determines the resilience of a system, and its risk of failure. Social capital influences actually both, the probability and size of damage. This became suddenly clear to me, when listening to one of the seminars of ETH Zurich's Risk Center. This Risk Center brings together experts in probability theory with experts in complexity and network theory. In that seminar, it was discussed that large disasters have an over-proportional impact on public opinion. That's why plane crashes and terror attacks matter a lot, while people seem to feel less threatened by everyday risks such as car accidents or deaths of smokers. In some sense, the conclusion was that "size matters," i.e. large disasters make people respond in an irrational, perhaps even panic way. However, having studied the phenomenon of panic for some time, I can say that well-educated and well-informed people don't easily panic, in particular if there is not the element of sudden shock, but enough time to cognitively process the event.

Therefore, I concluded: "on average, well-informed people don't overreact to a disaster, but instead they respond to the fact that there has been more damage than just the physical one—namely damage to the social capital." For example, a large-scale disaster often damages people's trust into the risk management of companies or public authorities, in particular when it was caused by unprofessional behavior or corruption. While people care about such things, no insurance company is covering this damage to the social capital.

My next conclusion was: we need to protect social capital in a similar way as we protect economic capital or our environment. Social capital can be damaged and exploited like our environment, but this should be prevented. For this, we must be able to measure social capital and quantify its value. Eventually, quantifying the value of our environment also helped to protect it better...

17.7 Trust and Power

To stress the importance of social capital, it is important to recall that the financial crises resulted from a loss of trust: banks did not trust other banks anymore and did not want to lend their money; customers did not trust their banks anymore and emptied their bank accounts; banks did not want to give loans to companies anymore; people did not want to invest in financial derivatives anymore, etc. In the end, the financial meltdown resulting from this amounted to an estimated 20 trillion US dollars. So, trust has a pretty high value, and when it gets lost, the economic losses are tremendous. To give another example: the recent loss of trust into US cloud storage companies after the NSA scandal was estimated to cause an economic loss of 35 billion US dollars, which is significant as well.

Trust is also the basis of power and legitimacy. One day I noticed that a deadly car accident in the city of Göttingen, Germany, caused by a mistake of the police, caused a large public outcry and demonstrations. This was the first instance, when I understood that public institutions can easily lose their public support, in other words: their social capital. This happens, if trust gets lost over something that the authorities should not have done according to the moral beliefs of the public.

I made the same observation in Zurich, Switzerland, where there were many complaints about the work of the migration office. During this time, the windows of the migration office were repeatedly smashed. However, when the office director was replaced, the problem disappeared. Furthermore, the London riots were triggered by an event, where the police shot a person without giving a sufficient justification to the public. And, in fact, riots in many other countries were triggered by events where public authorities were perceived to not have done a proper job. The Arab spring, for example, started in Tunisia, after Mohamed Bouazizi burnt himself, because of police corruption and improper treatment.

In other words: legitimacy and power result from doing the right thing from the perspective of the people. When people don't offer their idealistic or practical support anymore, authority and power are lost. While one may buy weapons and, with this, destructive power, constructive power depends on the trust and support of the people—otherwise they won't provide support, and this basically means that one hasn't got any power. Brutality does not create respect. It may create fear, but this can replace legitimacy only to a certain point. As the situation gets increasingly unacceptable, more and more people will lose their fear, start to resist the previously respected authorities actively or passively, or become ready to sacrifice their lives.

But even passive resistance will bring a country economically to a rest within just a few years or decades. This could, for example, be seen in the former German Democratic Republic. In conclusion, trust seems to be the only sustainable basis of power and social order. It must, therefore, concern us that, in many countries, politics and management are currently the professions with the lowest levels of reputation. In contrast, social professions that create public goods—firefighters, scientists, doctors, nurses and teachers—earn the highest levels of reputation.

17.8 Creating Social Capital

The interesting point about social capital is: if we understand how it comes about, then we can also create it. In fact, Social Information Technologies can open up new possibilities to produce social capital. The Social Mirror helps us to behave more beautifully, hence, to be more attractive and influential. The Social Adapter identifies opportunities for favorable interactions, which would otherwise be missed. The Social Adapter, furthermore, supports us in identifying opportunities and in performing cooperative projects, while considering side effects and externalities. Therefore, Social Information Technologies are tools to help us succeed in social and other affairs. The resulting social capital can be used to produce economic success as well. I am ending this chapter with an insight of New York City's Chief Urban Designer, Alex Andros Washburn. Referring to the Central Park, he said: public space is where you build public trust, by bringing all sorts of people together. Thus, in order to build trust in the digital world, we must create information commons and participatory opportunities, where all people and stakeholders can meet and exchange information transparently and openly.

Chapter 18
Digitization 2.0: A New Game Begins

Dirk Helbing

The political utopias facilitated by the Internet and social media have faded. However, the true digital revolution is yet to come—one that combines prosperity, sustainability and peace.

Recently, technology visionary Elon Musk, head of Tesla and Space X, asked, "What if the world were a computer simulation?" Then, life would be like a game where, in order to win, we would have to learn how to creatively get to the next level! But what are the rules?

The challenges of the game should really be known to all of us. More than 40 years ago, the "Limits to Growth" study, commissioned by the Club of Rome, found that a world with limited resources would inevitably run into an economic and population collapse.

Irrespective of the model parameters used in the computer simulations of the world's future, it always ended in disaster. As a result, decision-makers seemingly panicked. It was believed that billions of people would have to die. Step by step, a "fight of everyone against everyone" began. The prevailing strategy was to bring as many resources under control as possible. We decided to play "Monopoly". Consequently, the last decades of our planet were characterized by globalization and wars.

This article by Dirk Helbing was first published in June 2017 on ResearchGate under the URL https://www.researchgate.net/publication/317279118. It is an updated and expanded translation of an essay published by the NZZ on December 5, 2016, which is accessible under the URL https://www.nzz.ch/meinung/kommentare/digitalisierung-20-es-beginnt-ein-ganz-neues-spiel-ld.131433

D. Helbing (✉)
ETH Zurich, Zürich, Switzerland

TU Delft, Delft, Netherlands

Complexity Science Hub, Vienna, Austria
e-mail: dhelbing@ethz.ch

18.1 "Bread and Circuses"

Apparently, no one thought of changing the system of equations that governed our future, i.e., the way in which we organize our economy and society. What if the goal of the game was a different one, namely, to find new rules of the game so that the world's resources would be enough for everybody? Then we would have miserably failed! Indeed, fixing the problem would have been possible. If we had reduced our resource consumption by a miniscule 3% each year over the past 40 years, we would have a sustainable economy already. In fact, in the 1970s, an environmental movement emerged. There were car-free Sundays and people replacing Jute for plastic . . .

> What if the goal of the game is to find new rules of the game, so that the resources of the world are enough for everybody?

But this did not please the industrialists. From their perspective, citizens should continue to consume and not think about the future. The motto was bread and circuses for the people—distraction from the imminent apocalypse and the end of the world. Politics and industry would take care of everything. We just had to let them do what they believed had to be done. And we as a society trusted them—we cleared their way. . ..

The economic credo of the time was: "If problems are just large enough, there are enough incentive for engineers to invent a technical solution that can then be scaled up globally." In this way, problems could never get serious. For this plan to work out, industry should be restricted as little as possible. Therefore, neo-liberalism spread. It was claimed that, when resources become scarce, "economies of scale" were needed more than anything else, i.e., efficient global-scale production. Oligopolies and monopolies were the result.

New methods of energy and food production were developed, such as nuclear energy and genetically modified food. Also, the gross domestic product increased impressively. However, the energy consumption per person increased, and with the further spread of the oil industry, the world population grew by 2–3 billion people. Furthermore, even though petroleum companies knew about the climate impact of their business already in the sixties, political action was delayed for half a century by fueling scientific and public controversy. In summary, despite increased efficiency, the consumption of resources kept growing. In the end, consumers were blamed for this, even though the masses were influenced to buy many products they did not want or need, using giant advertising budgets and marketing campaigns.

18.2 Total Transformation of the Economy

Then followed the Paris climate agreement and countries had to admit that the efforts of big businesses had not been enough to solve the world's existential problems. Even among the Rockefellers, voices were heard that the petroleum business was immoral. Europe thereafter committed itself to reduce its CO_2 emissions by 40% in 15 years. Today, however, most of our economy—heating, transport and logistics, plastic and fertilizer production—is still built on coal, gas and oil. Hence, we need nothing less than a total restructuring of our economy. The alternative would be a drastic population reduction in the world. Our refusal to produce and live sustainably has become a question of life and death.

However, the solution to the impending resource crisis is neither to depopulate the world nor to impoverish the masses through a creeping financial, economic, and debt crisis. In contrast, a combination of a circular economy and a sharing economy would help. Within our lifetime, each of us produces 50 tons of waste—including several cars, TV sets, computers, mobile phones, all sorts of furniture and much more. This swath of resources would, in principle, be enough for at least five people if we finally replaced linear supply chains by circular material flows. While this cannot be efficiently achieved by regulation, a new, digital, financial and economic system could facilitate this by creating new kinds of market forces. We will discuss this later. In the meantime, the inequality in the world has reached an extent that, according to the OECD, WEF, and IMF, strangles economic development,[1] and—I would like to add—political, social, and cultural evolution certainly, too. In many European countries one can already see this clearly. For instance, although the central banks are pumping trillions of dollars into the financial markets (by Quantitative Easing), the global economy has not succeeded in recovering fully.

A world in which one percent of humanity controls as much wealth as the other 99 percent will never be stable. Barack Obama.

The world pays a high price for the fact that the monetary and financial system was clearly forgotten in the democratization of our society. Today, money creation is in the hands of very few people. The FED, for example, is 100% private, the European Central Bank is private in part, and commercial banks are private anyway. This leads to a serious conflict of interest, as private interests are standing above public interests. What is worse: those who had the resources and power did not solve the world's problems. It is therefore time for us to admit that society is suffering from the political and feudal structures that still determine the world in many ways. However, it makes no sense to keep a financial and money system alive that no longer meets the demands of the modern world and requires billions to die. There is no legitimacy for such a system. High time to change it!

---

[1]Inequality hurts economic growth, finds OECD research, see http://www.oecd.org/newsroom/inequality-hurts-economic-growth.htm (accessed July 15, 2017); Rising inequality threatens world economy, says WEF, see https://www.theguardian.com/business/2017/jan/11/inequality-world-economy-wef-brexit-donald-trump-world-economic-forum-risk-report; IMF study finds inequality is damaging to economic growth, see https://www.theguardian.com/business/2014/feb/26/imf-inequality-economic-growth

18.3 "Democratic Capitalism"

One cannot put the interests of a few hundred people over those of mankind. As long as the mechanism of money-making does not benefit everyone equally, the fate of the world will hardly change for the better. But there are alternatives! The failed approach of pumping trillions into the economy from the top could be replaced by a new approach, where the money is fed in from the bottom, i.e., via the bank accounts of the citizens.[2] At least a part of this money should be paid out as an investment premium. Everyone should invest it into projects, i.e. distribute it to people with good ideas, and to people who engage for social or environmental matters.[3] Then the money would flow where the best ideas and engagement are. I think it would be the logical next step to go from venture capitalism, to crowdfunding, to participatory budgeting, and finally to crowd funding for all. Such an approach would marry together our two most successful systems—democracy and capitalism—and replace today's "market-conform democracy", where capitalism threatens to destroy democracy, by "democratic capitalism". According to the constitution, "all men are equal before the law." It is time to demand true equality of opportunity, also when it comes to money creation!

In this new democratic capitalism we would continue to earn an individual salary, depending on what we do for business, society, or the environment. But we could focus on what we consider to be important and right. Combined with "open innovation", the rate of inventions, innovations and investments would be accelerated in all areas of life, from the improvement of neighborhoods to the investment in new technologies. Isn't this what we need in order to allow society to master its challenges in a successful and timely manner? Couldn't we now reach the next level of our economic, social, and cultural evolution?

18.4 Behavioral and Social Control?

If we continue as before, things couldn't end well. We would surely end in disaster, as crises, wars, terrorism, mass migration and populism herold. We must take a new path. In many countries, consumers are already lacking purchasing power, and many companies are no longer able to profitably sell their products. As a result of both, citizens and entrepreneurs are suffering. Banks are in danger of collapsing, and many

---

[2]Note that I am not talking here about unconditional basic income, because it might eliminate healthy competition as well as creative and performance incentives.

[3]Towards Democratic Sustainability, see http://futurict.blogspot.de/2017/06/propositions-on-per spective-global.html; this is an extension of an idea described here: With this new system, scientists never have to write a grant application again, Science (April 13, 2017), http://www.sciencemag.org/ news/2017/04/new-system-scientists-never-have-write-grant-application-again; Dutch funding trial puts 'wisdom of the crowd' centre stage, see http://iopscience.iop.org/article/10.1088/2058-7058/ 29/8/14/pdf

states would bankrupt immediately if central banks would increase the interest rate only slightly. Hence, politicians must do what is demanded from them by those who own and run the central banks.

Inequality has again reached the level that existed before the French Revolution. Therefore, the current elites are afraid to lose control of world affairs. This is why digital instruments have been created to not only surveil all of us, but also steer our attention, opinions, feelings, decisions and actions—with personalized information.[4] "Cambridge Analytica" has shown how to do it in the Brexit and US campaign.[5] But it is not the first and most influential company to use the "mind control" techniques of "neuromarketing".[6] The personalized advertising of Google and Facebook, as well as personalized news feeds of Facebook and a wide range of Internet platforms—recommending anything from restaurants to hotels and holiday destinations to sex dates, partners and friends—manipulate increasingly many areas of our lives. Many of these decisions are influenced on the subconscious level and feel like our own. The success of these methods is based on psychological tricks and the collection of huge amounts of personal data, which makes us transparent and controllable.[7] The effectiveness of this new and computerized form of propaganda is further accentuated by "social bots".

Through profiling and big data analysis, the Internet knows us probably better than us, including our weaknesses and our diseases. Most of this information was collected without our knowledge and our consent. As was recently revealed, so-called "cookies" and "supercookies" collect all our clicks when using the Internet. It is now practically impossible for the normal consumer to use the Internet without being digitally exposed. By now, much more information has been collected about each of us than the Stasi or secret services of totalitarian states ever had. How long can this go well? A lot of money is made with selling our data. Our privacy, which should have been constitutionally protected by the state, has become a commodity of surveillance capitalism. We are all the focus of interest—not just terrorists.

---

[4]D. Helbing et al. Will democracy survive Big Data and Artificial Intelligence. Scientific American (February 25, 2017), see https://www.scientificamerican.com/article/will-democracy-survive-big-data-and-artificial-intelligence/ (reprinted with permission).

[5]H. Grassegger and M. Krogerus: The data that turned the world upside down, see https://motherboard.vice.com/en_us/article/mg9vvn/how-our-likes-helped-trump-win; The great British Brexit robbery: how our democracy was hijacked, https://www.theguardian.com/technology/2017/may/07/the-great-british-brexit-robbery-hijacked-democracy

[6]Mind control: The advent of neuroscience in marketing, see https://www.theguardian.com/media-network/media-network-blog/2012/mar/23/mind-control-advent-of-neuroscience

[7]Diese Firma weiss, was Sie denken [This company knows what you think], see http://www.tagesanzeiger.ch/ausland/amerika/diese-firma-weiss-was-sie-denken/story/25805157

18.5 "Digital Judgment Day"

Representatives of the first digitization phase—let's call it the "digitization 1.0"—often argue: If you only had enough data, the truth would reveal itself, even without theory and science. Superintelligent systems would understand the world and tell us how to optimize it. Then one would only have to do what the data tell us to do. The best kind of data-driven society would, therefore, be a "benevolent dictatorship". In view of the various crises and challenges of this world—such as climate change and its many victims—it is claimed that "the end justifies the means". In bad times, it would be necessary to limit human rights and, if necessary, even to sacrifice democracy. Is this really what we have to do, or has politics been deceived? Not to mention the public, who did not know anything about all this at all . . .

It is often argued that the world's problems are a consequence of the irrational behaviour of selfish citizens. They would destroy the environment with their consumption behaviour.[8] Therefore, one would have to correct their "misbehaviour" and steer their decisions. Business and politics would take care of this using personalized information ("big nudging"), personalized prices and a "Citizen Score". The Citizen Score is a number measuring the "value" or "usefulness" of a citizen from the point of view of whoever rules the country.[9] In the future, it would decide which products and services we get, what jobs, rights, and credit conditions. Everything that we and our friends do or don't do would give plus or minus points. Such a system is already tested in China, but not only there. A similar "Karma Police" program was revealed in Great Britain as well, run by the secret service.[10]

To determine the Citizen Score, an algorithm analyses your clicks on the Internet, including the videos you watch and the music you hear. Importantly, in case of resource shortages, the Citizen Score would decide who will receive what kinds of resources and services and who won't. For some, this may be very bad news. One may compare the approach with a data-driven "Judgment Day", run by an Artificial Intelligence system. This system is ready to use. During the next disaster or crisis, the system could be turned on, and then, the above described totalitarian, neo-feudalistic system may persist for decades. Self-determination, democracy and human rights would largely be lost. In other words, democracy is threatened by a possible power grab by a small elite, which is being justified by the world's problems caused by us.

---

[8]R. H. Thaler, Misbehaving (W.W. Norton & Co, 2014), see also https://www.youtube.com/watch?v=K2alOrZ7A6Y; Für gute Verbraucherpolitik sorgen [Taking care of good consumer policy], see http://www.ulrich-kelber.de/medien/doks/20120424_Fr-gute-Verbraucherpolitik-sorgen.pdf, page 13ff.

[9]Big Data, meet Big Brother: China invents the digital totalitarian state, see https://www.economist.com/news/briefing/21711902-worrying-implications-its-social-credit-project-china-invents-digital-totalitarian

[10]Profiled: From radio to porn, British spies track Web users' online identities, see https://theintercept.com/2015/09/25/gchq-radio-porn-spies-track-web-users-online-identities/

Experts agree that, what has become technically possible by combining Big Data with Artificial Intelligence, Smart Devices, the Internet of Things and Quantum Computers, exceeds the scenarios described in George Orwell's "1984" and Aldous Huxley's "Brave New World" exceedingly so. Therefore, several people—including the previous president of the European Parliament Martin Schulz—recently demanded that we fight against technological totalitarianism. The former US President Barack Obama also warned us that liberal democracies are under attack by forces that ignore science and facts—forces so powerful that it would not be enough to put a megaphone into the hands the people. In the meantime, we have seen that he was right. Public opinion is now increasingly controlled by intelligent computer programs such as social bots—and so we have arrived in a "post fact society" in which fake news increasingly determine the public discourse.

18.6   "Deadly Danger"

Recently, Frank-Walter Steinmeier—when he was Minister of Foreign Affairs—called the "post fact society" created by modern manipulation techniques a "deadly danger" for democracy. And Elon Musk called Artificial Intelligence the perhaps greatest threat to humanity, possibly more dangerous than nuclear bombs. Therefore, he invested one billion dollars to establish the OpenAI Initiative with the aim of ensuring that Artificial Intelligence will be democratically used as "an extension of individual human wills and, in the spirit of liberty, as broadly and evenly distributed as possible". Also, a unique initiative of five large IT companies—IBM, Google, Amazon, Facebook, and Microsoft—was launched to ensure that Artificial Intelligence will be ethically used and destroy the "echo chambers", which are made responsible for increasing extremism and polarization. Finally, the initiative addresses the problem that people have been locked up in an informational filter bubble (a kind of "digital matrix"), which hinders our free, creative, and innovative thinking.

However, many things played out differently than expected. For decades, US strategist Zbigniew Brzezinski had propagated that the world was a great chess board, such that a kind of giant chess computer was built with the goal to control us all and win the game.[11] However, recently he had to bury the dream of American world supremacy. After Brexit and the outcome of the US elections it appears that the previous world order had come to an end. The old organizational structures of our world were increasingly breaking into pieces before our eyes. The USA—once the role model of the Western world—had exhausted itself in countless globalization wars. The financial system and infrastructure were crumbling. The society and its value system were breaking apart, and the US election campaign has turned the

---

[11]Z. Brzezinski, The Grand Chessboard—American Primacy and its Geostrategic Imperatives (Basic Books, 2016), see https://www.amazon.com/Grand-Chessboard-American-Geostrategic-Imperatives/dp/046509435X/

country into a battlefield. In the meantime, the Shanghai Cooperation Organization was becoming a superpower, and the BRICS countries (Brazil, Russia, India, China and South Africa) were calling for new financial and economic structures. They were less and less willing to provide material goods for worthless paper money...

> There will be again competition for the most innovative system. In the end, the best ideas will win, not money or power.

The Paris Climate Change Agreement will also lead to fundamental changes.[12] There will again be competition for the most innovative system. In the end, the best ideas will win, not money or power. To be more creative, we will need more scientific, economic, and political freedoms and opportunities than today. Thanks to a new financial system, environmentally and socially responsible behaviour will be rewarded. A sharing economy and a circular economy will enable more people to enjoy a high quality of life with fewer resources. New energy systems based on decentralized energy production will be established. But there is still a long way to go. All of us will have to make an effort to get our society to the next level together.

18.7 A Digital Ecosystem

First, we must learn that our reality is based on co-creation and co-evolution. We need collective awareness that our well-being is dependent on the environment and our fellow human beings, and that it is best for us to cooperate. Digital technologies can help us get it all started. When we use them properly, we will experience a "golden age"—a new era of prosperity, sustainability, and peace.

But how do we get into this new age? The digital revolution is instrumental for this. In the first phase of the digital transformation, Big Data, and Artificial Intelligence were used to create central information and control systems. But since the US election, the Silicon Valley is in a crisis. The technocratic visions of automated Smart Cities and Smart Nations have not kept their promises. "More prosperity for all" has not been accomplished. A much-noted Open Letter on the Digital Economy[13] calls for a re-orientation, something like a "digital ecosystem" that can benefit us all, not just a few large companies.

In this way, we enter the second phase of the digital transformation, the "digitization 2.0". It will be characterized by principles such as co-creation, co-evolution, collective intelligence, self-organization, and self-regulation. Coordination will be more important than control, and empowerment more important than power. This evolution will create a giant sharing economy, in which everyone can participate with their own ideas, products and services. Reputation and reciprocity, as well as

---

[12]Paris Agreement, see https://en.wikipedia.org/wiki/Paris_Agreement and http://unfccc.int/paris_agreement/items/9485.php

[13]Open Letter on the Digital Economy, see http://openletteronthedigitaleconomy.org/

data portability and interoperability, will be important functional principles. They will enable efficient exchange of ideas and resources as well as combinatorial innovation, and thus an explosion of creative and economic possibilities.

It will be seen that the digital economy is completely different from the material one. The latter is characterized by the competition for limited resources. The digital world, in contrast, benefits from the sharing of non-material resources, which are unlimited in principle.

> With digital technologies, it becomes easy to share and recycle resources.

It is now necessary to learn how to play this new game. It is a cooperative game, not the "Monopoly" of the old, material economy. The principle of ownership is increasingly replaced by the principles of use, access, and sharing. In addition, it suddenly becomes possible that the limited resources of the material world are sufficient for all. We just need to learn to recycle and share resources. We need a circular economy rather than linear supply chains, where fresh resources are used to produce consumer goods, which are finally thrown away.

But how can we bring the new system forward? The digitization 2.0 will come with three closely intertwined transformations: the digital, the ecological, and the financial transformation. The Internet of Things and Blockchain technologies are the technological drivers of the financial transformation. By linking them together and taking the science of complex systems into account, it is possible to manage complex systems such as society or nature in real time. And as AI-driven automation in the wake of the digital transformation takes over 50% of today's tasks previously performed by us, we can concentrate ourselves on the issues that have been in part neglected so far, including environmental and social issues.

18.8   New Incentive System

In other words, we have to re-invent half our economy and build an ecological, digital economy that consumes much less coal, gas, oil, and other resources. Sustainability could be reached with a new, differentiated incentive system—a socio-ecological financial system that is liberal, efficient and democratic at the same time. How can this be achieved? The financial system is essentially a coordination mechanism, which decides who receives how much of what resource at what price. But there could be a myriad of better coordination systems. Instead of managing society with a complicated tax system with 1–2 years delay, the Internet of Things will soon allow for real-time taxation information. This can be set up in such a way that the values of society are built into the system ("values by design").

The effects of our actions, including our "externalities", can now be measured at low cost: noise, stress, CO_2, emissions, waste, etc., but also desired outcomes such as job creation, social cooperation, education, health, and the reuse of resources. These would receive a price or value in the socio-ecological "finance system 4.0". With the addition of numerous new currencies, existing alongside today's one-dimensional monetary system, one could increase the desired effects and activities and reduce unwanted ones. Social and ecological commitment would no longer be expensive—it would pay off. With such an approach, a circular economy would basically emerge by itself, driven by new market forces rather than a digital command economy. Numerous regulations could be replaced by measurement processes and participatory (subsidiary) pricing processes. Through a hierarchy of incentive systems, one could promote local commitment to achieve global goals. The economy would become efficient, and it could benefit all: citizens, banks and businesses. In the interest of digital democracy and collective intelligence, the socio-ecological financial system would be jointly managed by representatives of the economy, politics, science, and the general public.

In addition, the socio-ecological finance system could be designed in such a way that it would automatically generate taxes to pay for public goods and infrastructures. By means of the differentiated, multi-dimensional incentive system, once could manage complex systems much better, and even build self-organizing or self-regulating systems. The externalities underlying this incentive system would be measured in a crowd-sourced way, using sensors in smartphones and the Internet of Things. By sharing the measurements and making the data available to all, one could earn different kinds of money—money that would be fed into the economy on the bottom, from where it would eventually flow to the top. Even without a redistribution of money and wealth, we could all benefit, simply by organizing the use of resources much better.

Here is an example: With its self-driving cars, Google would soon like to offer "transport as a service". The company believes that about 15% of today's vehicles could cover our mobility requirements. This means we would need a lot less steel, rubber, glass and other materials, fewer garages, less parking lots, etc. It would also be more comfortable for us. You would just tell your smart device that you need a vehicle in 5 minutes, and it would bring you from door to door, then move on to the next customer. It might be even cheaper than using a car today...

To take a new path, we first need to break up the shackles of the old age.

Why don't we start this now? Europe missed the train of the digitization 1.0. So what? Let us be world champions of the digitization 2.0! We could be pioneers in building digital democracy, the socio-ecological finance system, and democratic capitalism. When robots soon produce the goods we need to live, we can spend our time with creative and social activities, with learning and environmental projects. Digital assistants could help us in all situations. Personal Artificial Intelligence systems, acting in our interest, would help us manage our personal data and support our informational self-determination. With spoken instructions or even steered by our thoughts, we could create and experience new worlds and Virtual Realities.

However, we are not there yet. We first have to free ourselves from the fetters of the old age before we can take a new path. Which one would we choose? It is time for a public debate where we want to go in the digital age—and for investments into the future rather than in the renovation of the broken past. Let's do this now!